Rotating Machinery:

Practical Solutions to Unbalance and Misalignment

Rotating Machinery:
Practical Solutions to Unbalance and Misalignment

by Robert B. McMillan, PE, CEM

Routledge
Taylor & Francis Group

LONDON AND NEW YORK

Published 2020 by River Publishers

River Publishers

Alsbjergvej 10, 9260 Gistrup, Denmark

www.riverpublishers.com

Distributed exclusively by Routledge

4 Park Square, Milton Park, Abingdon, Oxon OX14 4RN

605 Third Avenue, New York, NY 10017, USA

Library of Congress Cataloging-in-Publication Data

McMillan, Robert B., 1944-

Rotating machinery: practical solutions to unbalance and misalignment/by Robert B. McMillan.

p. cm.

I. Includes index.

ISBN 978-0-8247-5052-7 (print) ISBN 978-8-7702-2243-3 (electronic)

1. Balancing of machinery. 2. Machinery--Vibration. I. Title.

TJ177.M42 2004

621.8'11--dc22

2003060258

Rotating machinery: practical solutions to unbalance and misalignment/ McMillan, Robert B.

First published by Fairmont Press in 2004.

Routledge is an imprint of the Taylor & Francis Group, an informa business

0-88173-466-7 (The Fairmont Press, Inc.)

978-0-8247-5052-7 (print)

978-8-7702-2243-3 (online)

978-1-0031-5114-2 (ebook master)

Table of Contents

Preface

I t is the intent of this book to present both a theoretical and the practical understanding of unbalance and misalignment in rotating equipment. These two conditions account for the vast majority of problems with rotating equipment encountered in the "real world." Numerous examples and solutions are inserted to assist in understanding the various concepts.

Chapter 1 discusses vibration to provide a fundamental understanding of its characteristics, which are used to determine the operational integrity of rotating machinery. A section deals with determining the criticality of equipment to provide a set of guidelines as to which equipment needs to be more closely monitored. Finally, the chapter ends with a discussion of how to take and record field vibration measurements.

Chapter 2 details the relationships between the various vibration characteristics to gain an understanding of the forces generated within operating machinery when conditions of unbalance and misalignment are present.

Chapter 3 deals with resonance and beat frequencies, and details their sources and cures.

Chapter 4 discusses the various forms of unbalance that can be present in rotating machinery. Many of these sources can be corrected during inspections and when performing sound maintenance practices.

Chapter 5 presents both the single plane and dual plane methods of balancing rotating equipment. A discussion of which method to employ is also presented. Also presented are methods of determining the correct balancing weights and how to resolve them to locations where they can be attached.

Chapter 6 deals with the three circle method of balancing slow speed fans, where phase angles are difficult to obtain.

Chapter 7 discusses the various types of misalignment and how to detect them through vibration analysis.

Chapter 8 outlines an advanced rim and face method of precision alignment. Both graphical and calculator solutions are presented. Thermal growth, spool pieces, equipment trains, vertical pump equipment with driveshafts and the catenary effect of long shafts are presented.

Chapter 9 presents the reverse indicator method of alignment with both graphical and calculator methods. Again, examples are used to fortify the learning experience.

The appendix section has several useful tables and charts to assist the reader in solutions to other problems. In addition, a complete set of procedures for fan maintenance is included as guidelines for developing maintenance procedures for other equipment.

I sincerely hope the readers of this book will find it useful in solving the everyday problems within their facilities. Your comments, suggestions, and criticism will be highly regarded.

Robert B. McMillan, PE, CEM
El Paso, Texas
June, 2003

Chapter 1

Introduction to Vibrations

Vibration is the back and forth or repetitive motion of an object from its point of rest. Simple vibration can be illustrated with a mass suspended by a constant force spring, as shown in Figure 1-1.

When a force is applied to the mass, it stretches the spring and moves the weight to the lower limit. When the force is removed, the stored energy in the spring causes the weight to move upward through the position of rest to its upper limit. Here, the mass stops and reverses direction traveling back through the position of rest to the lower limit. In a friction-free system the mass would continue this motion indefinitely.

In real situations, there is always some form of friction or dampening which causes the mass to not quite reach the upper and lower limits, and gradually lowers the motion until the mass returns to the position of rest.

However, if the force were applied again, the mass would again move to its lower limit. By successively adding and removing the force, the mass would continue to vibrate. Although this is a very simplified example, all machine parts exhibit the same basic characteristics.

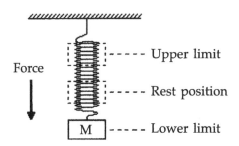

Figure 1-1. Simple Vibratory Motion

1

It is important to note that an external force is required to keep the mass in motion. All real systems are damped, that is they will gradually come to their rest position after several cycles of motion, unless acted upon by an external force. When this vibratory motion is viewed with respect to time, it is seen as a sine wave, as illustrated in Figure 1-2. Note that continued motion would only repeat the sine wave as shown.

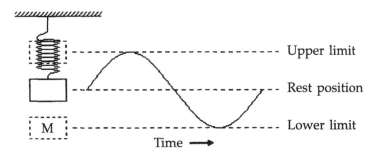

Figure 1-2. Generation of a Sine Wave

Some of the characteristics of this vibratory motion are **PERIOD, FREQUENCY, DISPLACEMENT, VELOCITY, ACCELERATION, AMPLITUDE** and **PHASE**. The sign wave shown in Figure 1-2 represents one complete cycle, that is the weight moved from its rest position to its upper limit back through its rest position to its lower limit and then returned to its rest position. Continued vibration of this spring mass system would only repeat the characteristics shown in this single cycle.

PERIOD

The period of the vibration, represented by the letter T, is the time required to complete one oscillation. That is the total time required for the mass to move from the rest position to the upper limit, back through the rest position to the lower limit, and return to the rest position.

FREQUENCY

Frequency is the number of complete oscillations completed in a unit time, or simply expressed as **1/T**. Generally it is measured in either cycles per second or cycles per minute.

In dealing with machine vibrations, it is usually desirable to express frequency in terms of cycles per minute, since we measure the rotational speed of machinery in revolutions per minute. This allows examination of the vibration frequency in terms of multiples of the rotational speed. The two times rotational speed is known as the second harmonic, and the three times rotational speed is the third harmonic.

Rotational speed is also known as the fundamental frequency. Figure 1-3 displays some of the characteristics of a simple vibration.

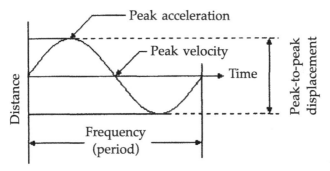

Figure 1-3. Characteristics of Simple Vibratory Motion

DISPLACEMENT

Referring to Figure 1-3, the displacement of the mass can be seen. The peak-to-peak distance is measured from the upper limit to the lower limit, and is usually measured in mils or thousandths of an inch (.001 inch). In the metric system, the displacement is measured in microns or millionths of a meter (.000001 meter or .001 millimeter).

NOTE: The Appendix section contains some conversion factors for the metric and English systems of measurement.

VELOCITY

The velocity of a vibrating object is continually changing. At the upper and lower limits, the object stops and reverses its direction of travel, thus its velocity at these two points is zero.

While passing through the neutral or position of rest, the velocity is at its maximum. Since the velocity is continually changing with respect to time, the peak or maximum velocity is always measured and commonly expressed in inches-per-second peak.

In the metric system, vibration velocity is measured in meters per second peak. When expressing the vibration characteristic in terms of velocity, both the displacement and frequency are considered. Remember velocity is inches-per-second, and thus the displacement [inches] and the frequency [times per second] are considered.

ACCELERATION

Again referring to Figure 1-3, the peak or maximum acceleration is shown to be at both the upper and lower limits of the mass travel. Acceleration is the rate of change of velocity with respect to time. It can also be expressed as the rate of change in distance with respect to time, with respect to time. That is it is the change in displacement with respect to time squared. Normally, vibration acceleration is measured in terms of G's or number of times the normal force due to gravity. Gravitational force has been standardized as 32.1739 feet per second per second or 386.087 inches per second per second. In the metric system, the standard for gravity is 980.665 centimeters per second per second. Once again, acceleration is measured as G's peak.

Since the vibrating object must reverse course at the peak displacements, this is where the maximum acceleration occurs. Like velocity, acceleration is constantly changing, and the peak acceleration is usually measured.

The importance of measuring the vibration in terms of acceleration is best understood by examining Newton's first law of motion:

$$F = MA \tag{1.1}$$

Where:
- M = Mass in pounds (1 pound Mass = 32.1739 Pounds)
- A = Acceleration (feet per second per second)
- F = Pounds of Force

Note that the force of an object at rest is WG [its weight W times the acceleration of gravity G]. Since the standard acceleration of gravity is 32.1739 feet per second per second, mass is simply an object's weight times the acceleration of gravity.

Thus by measuring vibration in terms of acceleration, the force imparted by the vibration is also considered. It is this force that is important in determining the potential hazard to the machine and its components while undergoing vibration.

PHASE

Phase is the position of a vibrating object with respect to another object at a given point in time. In the simple example shown in Figure 1-4, two weights are vibrating 180 degrees out of phase with each other.

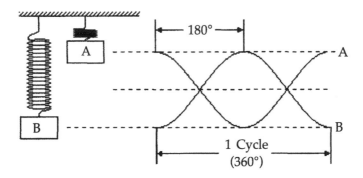

Figure 1-4. Phase Relationship 180° Out of Phase

Here, the phase angle is expressed in degrees, where there are 360 degrees in one cycle. Although both weights are vibrating in this example, it is possible to express a phase angle of a single vibrating weight with respect to a fixed object. Using the upper limit of motion as a reference point, the phase angle can be expressed in degrees from a fixed reference point. In Figure 1-4, weight A has a phase angle of zero degrees and weight B has a phase angle of 180 degrees.

It should also be noted that both vibrating weights have the same frequency, and thus will remain 180 degrees out of phase as long as they both are in motion. As will be seen later, this is not always the case. When the two vibrating weights have different frequencies, they will come in and out of phase with one another. This will result in what is known as a BEAT frequency. Generally, beat frequencies are minimal unless the two sources are within 20 degrees of each other. The beat frequency is the frequency of the two vibrating sources coming into and out of phase with one another.

In Figure 1-5, the two weights are vibrating 90 degrees out of phase with one another. In Figure 1-5, the two weights have the same frequency and will remain 90 degrees out of phase as long as they continue to vibrate. In Figure 1-6, the same two weights are shown vibrating with different frequencies. Note the resulting vibration when the two amplitudes are added to one another. Also

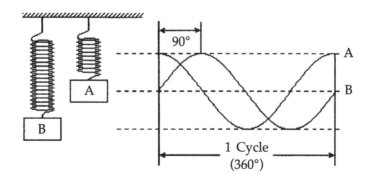

Figure 1-5. Phase Relationship 90° Out of Phase

note that this vibration will produce a beat frequency.

These characteristics of vibration are useful in determining the source of the vibration.

Normally, velocity is the preferred measurement of vibration for machine condition monitoring, because it considers both the magnitude and the frequency of the vibration. This is important in metals that can fail from fatigue. Fatigue failures are a function of the amount of stress applied, and the number of times it is applied.

Bending a coat hanger until it breaks is an example of a low cycle fatigue failure. In this case, a large stress was applied over a relatively low number of cycles. In most fatigue failures, the stress applied is considerably less; however, the number of cycles may exceed hundreds of millions. Consider an electric motor operating at 3,600 rpm. In one year, it rotates 1,893,456,000 times. It becomes obvious that a small stress applied that many times could lead to a failure.

Displacement measurements can be important, especially in low frequency vibrations on machines that have brittle components. That is, the stress that is applied is sufficient to snap the component. Many machines have cast iron frames or cases that are relatively brittle and are subject to failure from a single large stress.

Acceleration measurements are also important in that they

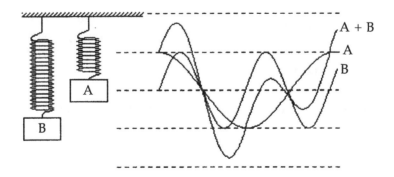

Figure 1-6. Two Sources with Different Frequencies

directly measure force. Excessive force can lead to improper lubrication in journal bearings, and result in failure. The dynamic force created by the vibration of a rotating member can directly cause bearing failure. Generally a machine can withstand up to eight times its designed static load before bearing failure occurs. However, overloads as little as 10% can cause damage over an extended period of time. Although this seems insignificant, it can be shown that small unbalances can easily create sufficient dynamic forces to overload the bearings.

HOOKE'S LAW

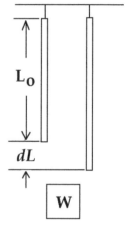

Referring to Figure 1-7, a metal bar is stretched a distance *dL*, by applying a weight W.

Graphing the amount of stretch versus the amount of applied weight produces a stress-strain curve as shown in Figure 1-8. In the elastic zone, if the weight is doubled, the stretch is also doubled. When the weight is removed, the bar returns to its original length.

Hooke's Law states that the amount of stretch, or elongation, is proportional to the applied force.

Stated as an equation:

Figure 1-7. Bar being Stretched with a Weight

$$F = (\text{constant})(dL) \qquad (1.2)$$

However, if the force applied is too large, the metal bar will reach its yield point, and when the force is removed, the bar will have a permanent elongation. The bar was stretched beyond its elastic limit.

Figure 1-9 illustrates the permanent offset caused by applying a force beyond a material's elastic limit. Note the yield point

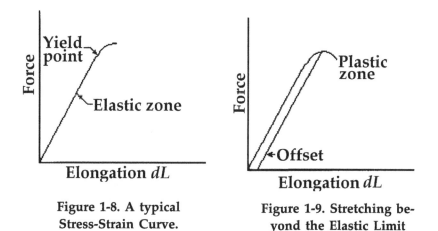

Figure 1-8. A typical
Stress-Strain Curve.

Figure 1-9. Stretching beyond the Elastic Limit

is where the material changes from elastic to plastic. As the force is continually increased, the cross-sectional area of the material is reduced and thus the curve begins to reverse. Eventually, the force per unit area is too great and the material fails.

Stress is defined as the force per unit area or:

$$\text{Stress} = F/A \tag{1.3}$$

Strain is defined as the elongation divided by the original length or:

$$\text{Strain} = dL/L_o \tag{1.4}$$

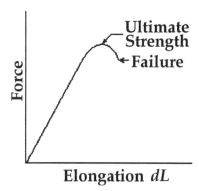

Figure 1-10. A Material being Stressed until Failure

For a simple mass spring system that vibrates within the elastic limits of the spring, Hooke's Law applies and is expressed as:

$$F = -kx \tag{1.5}$$

Where:
    $k =$ the spring constant
    $x =$ the distance from the rest position

The minus sign in this equation indicates that the force of the spring is acting opposite to the direction in which the spring is stretched.

If the spring were stretched from $x = 0$ to $x = x_0$ work is done which is equal to the average force times the distance moved. The average force is half the maximum force and the distance is x_0. Therefore, the energy in the spring is:

$$PE = (1/2k \; x_0)(x_0) = 1/2k \; x_0^2 \tag{1.6}$$

The sum of the potential energy [PE] and the kinetic energy [KE] must remain constant and equal to the maximum potential energy, $1/2k \; x_0^2$. Kinetic energy is the energy due to motion and is expressed as $1/2 \; mv^2$. Thus expressed in equation form:

$$PE + KE = 1/2k \; x_0^2 \tag{1.7}$$

Substituting:

$$1/2k \; x^2 + 1/2 \; mv^2 = 1/2k \; x_0^2 \tag{1.8}$$

Or:

$$mv^2 = k(x_0^2 - x^2) \tag{1.9}$$

Now substituting from Equation (1.1) into Equation (1.5):

$$-kx = ma \tag{1.10}$$

Or:

$$a = -(k/m)x \tag{1.11}$$

Equations 1.9 and 1.11 are common forms for describing simple harmonic motion. This is best illustrated by an example.

Example 1.1
A spring is stretched 1 foot with a force of 2 pounds. If a 6-pound weight is displaced 4 feet from the position of rest, (a) what is the maximum velocity, (b) what is the maximum acceleration, and (c) what is the velocity and acceleration when the distance is 2 feet?

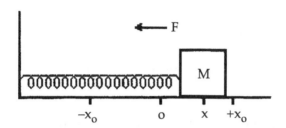

Figure 1-11. Mass Spring System for Example 1.1

Step 1
First, the spring constant must be determined. Using Hooke's Law,

$F = -kx$ or -2 lbs. $= (-k)(1)$ft. Therefore $k = 2$ lbs./ft.
The maximum potential energy is $1/2 \, k \, x_0^2$, or
$1/2k \, x_0^2 = 1/2(2 \text{ lbs./ft.})(16 \text{ ft.}^2) = 16$ ft.-lbs.

Step 2
Since the highest velocity occurs when $x = 0$, refer to Figure 1-3, using Equation (1.9) and substituting the provided data from the example yields:

$$\frac{6lbs}{32.2ft/sec.}v^2max =\left(2 \text{ lbs./ft.}\right)\left(16 \text{ ft}^2\right)$$

Note: Mass is the weight divided by the acceleration of gravity (32.2 ft/sec.^{2}). In this example, the mass is 6/32.2 or 0.1863 lbm.

(a)   V max = 13.1 ft./sec.

The maximum acceleration occurs when $x = x_0$, using Equation 1.11:

(b)   a max = $(2/0.1863)(4)$ = 42.9 ft./sec.^{2}

When the displacement equals 2 feet, using Equation 1.9

(c) $V^2 = k(x_0^2 - x^2)/m = (2)(16 - 4)/(0.1863)$
Therefore V = 11.35 ft./sec.
Using Equation 1.11

(d) $a = (k/m) x = (2/0.1863)(2)$ = 21.47 ft./sec.^{2}

VIBRATION AS A DIAGNOSTIC TOOL

In the following chapters, there will be a more detailed look at vibration as a diagnostic tool for determining unbalance and misalignment in rotating machinery. In addition, beat frequencies and natural or harmonic frequencies will be discussed in detail. The mode shape for objects vibrating in their first, second, and third harmonic frequencies will also be presented. A great deal of emphasis is placed on determining what is taking place even if you do not have access to a vibration analyzer.

The characteristic of a machine's vibration(s) can be used to identify specific problems. There are numerous causes of vibration in machines, but about 90% of all problems are due to unbalance or misalignment. Some of the other sources of vibration are:

- Mechanical looseness
- Bad bearings (anti-friction type)
- Bent shafts
- Aerodynamic forces
- Worn, damaged or eccentric gears
- Bad drive belts or chains
- Hydraulic forces
- Electromagnetic forces
- Resonance
- Rubbing

The characteristics of vibration due to unbalance and misalignment are presented in the following chapters. These include the frequency and the plane in which the vibration is present. These simple measurements can be used to determine the conditions of unbalance and misalignment.

In general, the vibration will exist in the radial direction, axial direction, or both. The radial direction is usually broken up into the vertical and horizontal planes to better describe the characteristics of the vibration.

Figure 1-12 illustrates the two basic directions of vibration of a machine shaft. Axial vibration occurs along the axis of the shaft, while radial vibration acts outwardly from the center of the shaft. It should be noted that the radial vibration may not be in the same direction at opposite ends of the shaft. One end may be moving vertically upward while the opposite end may be moving horizontally to the left at the same moment in time. This is known as the phase angle relation and will be useful in determining the cause of the vibration.

Figure 1-12. Direction of Vibration

Phase angle relations as well as the direction of vibration are discussed in the chapters on unbalance and misalignment, to assist in identifying the source of the vibration.

Figure 1-13 shows the two directions of vibration with the radial vibration broken down into the vertical and horizontal planes.

Although a vibration analyzer is required to identify phase angle relations and to perform field or shop balancing of machine parts, many causes of vibration can be determined through inspections and proper assembly techniques.

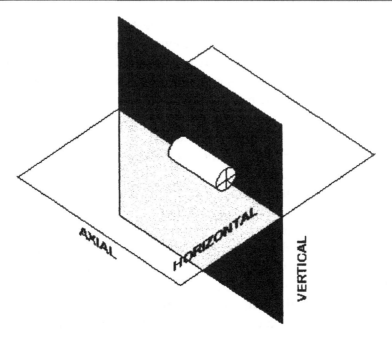

Figure 1-13. Direction of Vibratory Motion

The first step in the analysis process should be the recording of all machine information and drawing a sketch of the machine.

In figure 1-14, the basic data for a simple fan are recorded. It is important to know as many facts about a machine as possible. In this case, there are eight fan blades, which at 1,725 rpm would exhibit a first order aerodynamic problem at the blade pass frequency of 8 × 1,725 or 13,800 cycles per minute (cpm).

In analyzing the machine above, the frequency of 1,725 cpm

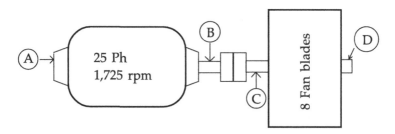

Figure 1-14. A Basic Sketch of a Fan to be Analyzed

(1 × rpm) will be important in determining both conditions of unbalance and misalignment. Two and three times operating speed should also be recorded. Note that a vibration analyzer with a filter is required to isolate these frequencies.

A filter out reading should always be recorded, to assure the filter in readings capture all the frequencies that sum to the filter out reading. In addition, the filter out reading is used to determine the severity of the vibration.

Prior to providing another example, the criticality of equipment needs to be discussed. There are certain pieces of equipment within a facility that demand more attention than others.

The amount of effort spent on any given piece of equipment will be directly dependent upon its criticality factor. Although each facility will need to build criteria of their own, the following illustrates how to categorize equipment.

IDENTIFYING CRITICAL EQUIPMENT

When surveying a facility to determine the extent of a predictive maintenance program to be implemented on an on-going basis, one important factor is determining the criticality of equipment. The prioritization of equipment will assure the proper focus on critical items and further assure that maintenance monies are wisely spent.

To evaluate each piece of equipment within the facility, the following chart can be used. Assign a value to each category based on the specific requirements of the individual company. Add the points scored by each piece of equipment and list them in descending order. You have now established a criticality list.

It is important that some base line be established as a reference point. That is, all equipment should be surveyed and their basic maintenance condition established. This provides the overall maintenance condition of the facility, and can assist in preparing a maintenance budget.

During the preliminary maintenance audit, mechanical and electrical condition is noted as well as performance data. As an example, centrifugal pumps should have a block off test run to determine the actual head produced. The voltage and amperes should be measured and the actual horsepower consumed calculated. These data should be plotted on the pump's performance curve and a record maintained.

A comprehensive listing of all equipment should be assembled, and any missing data should be requested from the manufacturer. Further, this will assist in determining what spare parts need to be carried in inventory, and what parts may be interchangeable.

EQUIPMENT IDENTIFICATION

No.	Category	Value	Yes/No
1	Hazardous Service	5.00	
2	Critical to Operations	3.50	
3	Chain Reactive	2.00	
4	Expensive to Repair	1.50	
5	One of a Kind	2.00	
6	Long, Tedious Repair	1.10	
7	Hard to find Parts	1.10	
Total Points			

The above form is filled out for each piece of equipment and then they are ranked by points. The equipment with the most points is the most critical and deserves the most attention.

The above categories are further defined below:

1. **Hazardous Service**
    This category of equipment can be a result of high temperature, high pressure, high velocity or the specific material being handled. Caustic, poisonous, corrosive, and flammable materials should be considered and assigned an appropriate value.

2.   **Critical to Operation**
     This category of equipment includes machines within the
     process that disrupt normal production. The cost for down
     time should be considered when assigning a value to this
     group of equipment. In general, this equipment has no
     backup or spare.

3.   **Chain Reactive**
     This category of equipment causes other machines or equip-
     ment to fail when it fails. As an example, if an oil pump
     supplying oil to the bearings of a large press were to fail, it
     could potentially cause failure of the press as well.

4.   **Expensive to Repair**
     This category of equipment covers machines that may be
     difficult to find parts for, parts are expensive, requires out-
     side labor, machining, special tools, etc.

5.   **One of a Kind (Exotic)**
     This category of equipment covers machines that are difficult
     to impossible to replace. This can be because of limited appli-
     cation, older equipment, foreign manufacture, or custom de-
     sign.

6.   **Long Tedious Repair**
     This category of equipment covers machines that require
     specialized labor and tools. The repair of this type of equip-
     ment may require contract labor.

7.   **Hard to find Parts**
     This category of equipment covers machines whose parts are
     not off-the-shelf items. Some of these parts may take weeks to
     obtain.

The following illustrates a form filled in for an acid pump:

NUMBER 2 ACID INJECTION PUMP

No.	Category	Value	Yes/No
1	Hazardous Service	5.00	Y
2	Critical to Operations	3.50	Y
3	Chain Reactive	2.00	N
4	Expensive to Repair	1.50	Y
5	One of a Kind	2.00	N
6	Long, Tedious Repair	1.10	N
7	Hard to find Parts	1.10	Y
Total Points		11.10	

This pump would score an 11.10 out of a possible 16.2 and should be considered as a critical piece of equipment. Analysis of this pump should be performed on a minimum of a 30-day basis, with the results trended and reviewed by maintenance supervisors. Consideration should be given to provide all possible predictive techniques for this piece of equipment.

RECORDING VIBRATION DATA

When a piece of equipment is analyzed, the data should be recorded and kept in a file for analysis and future review. Figure 1-15 illustrates a simple form for recording vibration data.

In Figure 1-15, the four basic points to measure vibration are labeled A through D. The area for recording data has three sections; A—axial, H—horizontal, and V—vertical. Readings should be taken and recorded for all the directions at the four identified points.

In addition, there are locations for both filter in and filter out. The filter in readings should be taken at multiples of the operating speed. In the event the majority of the vibration is not located, the entire spectrum should be scanned to determine the frequency of the vibration.

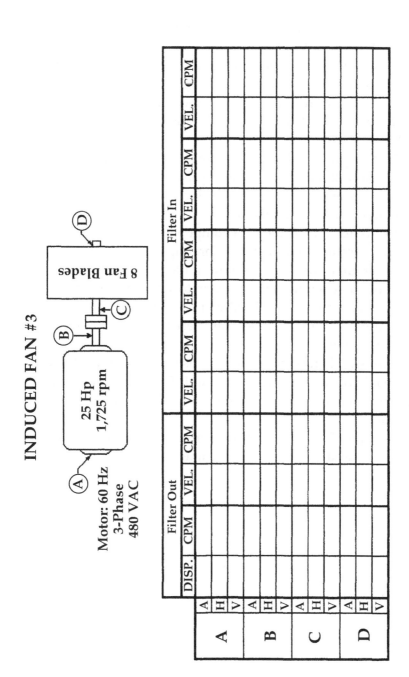

Figure 1-15. Form for Recording Vibration Data

Examples of analyzing vibration readings are presented in the chapters on unbalance and misalignment.

It is the primary intent of this book to allow technicians to perform simple field corrections and inspections to correct the majority of problems encountered. By feeling a vibrating part rotating at 1,750 rpm, using a screwdriver, the difference between 1 × rpm and several times rpm vibrations can easily be determined.

The 1 × rpm is felt as a vibration, where the higher frequencies feel like a tingling sensation. By placing the screwdriver on different parts and in different directions, the shape and direction of vibration can be determined.

Most often, a machine or a component will exhibit vibration in several directions and may be vibrating at several different frequencies simultaneously. It is important to understand the characteristics of these vibrations in order to identify the root cause. Various problems will exhibit different characteristics that will aid in evaluating the cause. These characteristics are discussed in the following chapters.

These simple techniques will be used to determine the cause and possible corrective actions required. There is no substitute for experience, and the more machines felt, the more confident the technician.

In Chapter 2, additional equations are presented to determine the forces caused by vibrating machine members.

Chapter 2

Developing the Vibration Equations

LIST OF SYMBOLS

Symbol	Subject	Units
A	Acceleration	Inches per second^2
A_{MAX}	Maximum acceleration	Inches per second^2
D	Peak-to-peak displacement	Mils
dx/dt	The derivative of displacement with respect to time	
dx^2/dy^2	The second derivative of displacement with respect to time	
f	Frequency	Hz
F	Frequency	Cycles per minute
G	Acceleration of gravity	Inches per second^2
π	Pi	3.14159
t	Time	Seconds
V	Vertical velocity	Inches per second
V_{MAX}	Maximum velocity	Inches per second
ω	Angle displaced	Degrees
X	Vertical displacement	Mils
X_o	Maximum displacement	Mils

In order to describe the characteristics of vibration, it is necessary to understand the basic trigonometric functions of sine and cosine. Since the vibratory motion repeats itself with respect to time, these functions are useful in describing the displacement, velocity, and acceleration associated with this type of motion.

THE SINE FUNCTION

To begin, we will construct a unit circle, that is a circle with a radius of 1 unit. It makes no difference what units are used—feet, inches, meters, miles etc.—because all functional results will be described as ratios of whichever unit is chosen. Figure 2-1 depicts a unit circle that is laid out from the three o'clock position, counter clockwise, and is so labeled.

THE UNIT CIRCLE

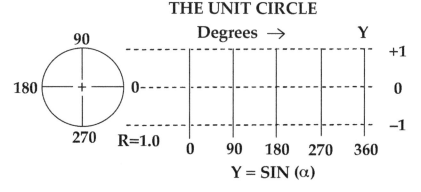

Figure 2-1. Laying Out a Unit Circle

To the right of the circle, three straight lines are drawn, one at zero, one a plus one (one radius), and the other at minus one radius. Also, along these lines are laid off 360 degrees, representing the number of degrees in the unit circle. In Figure 2-1, the Y direction is vertical, or up and down the page.

The X direction is laid off horizontally, or back and forth across the page. The X values will thus represent the number of degrees in the circle, while the Y values will represent the vertical distance that the end of the line representing the radius R travels from zero. Also note that if the radius were drawn from the center to the 12 o'clock position, the Y value would be plus one [+1], and if the radius were drawn from the center to the 6 o'clock position, the Y value would be minus one [–1].

By closely observing the unit circle in Figure 2-1, it can be seen if the degrees are increased from zero to 360, the end of the radius line will start at zero, increase to plus one, decrease to zero, continue to decrease to a minus one, and finally increase back to zero at 360 degrees. Note that 0 degrees and 360 degrees are the same point [3 o'clock].

This is the trigonometric sine function. The sine of any angle is defined as the Y value divided by the radius R or [Y/R]. In this special case where R =1, the sine of the angle is equal to the Y displacement. Note that the sine is zero for both angles 0 and 180, while the sine of angle 90 is plus one, and the sine of angle 270 is minus one.

In mathematical calculations, the sine of an angle is written as Sin(a), where a is the angle expressed in degrees or radians. Since this book uses degrees, all angles and their trigonometric functions are presented in degrees.

DEGREES OR RADIANS

There are 2π radians in a circle, and therefore one radian is equal to 180° degrees.

$$\text{Radians} = \text{Degrees} \times \text{pi}/180 \tag{2.1}$$

Or:

$$\text{Degrees} = \text{Radians} \times 180/\text{pi} \tag{2.2}$$

Thus the angle 27 degrees expressed in radians would be:

27 degrees = 27 × pi/180 radians, or .4712388980 radians.

When using calculators to obtain the sine or cosine of an angle, be sure to note whether the calculator calculates the functions in degrees or radians. If the sine function of 27 degrees was taken but the calculator was in radians, the answer would be .9564 instead of the correct answer of .4540. When 27 degrees is expressed in radians, it is .4712389. Calculating the sine of that number would yield the correct answer, namely .4540.

As the vertical distance for each degree is measured and plotted against the angle, a complete sine wave is generated. Figure 2-2 depicts a completed plot of the sine function for a revolution. That is starting at 3:00 o'clock (zero degrees) and proceeding counter clockwise back to the 3:00 o'clock position.

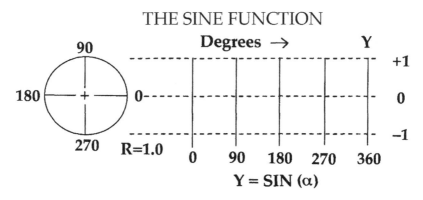

Figure 2-2. The Development of the Sine Function

In general, the sine of the angle describes the elevation of the end of the radius for each degree of rotation.

THE COSINE FUNCTION

The second trigonometric function to be defined is the cosine. Just as the sine function was laid out using a unit circle, so will the cosine be defined. The cosine of an angle is defined as the radius R divided by the X value. Note that the cosine function is zero at 90 and 270 degrees, plus one at 0 degrees, and minus one at 180 degrees.

The complete circle plot of the cosine function is illustrated in Figure 2-3. Note that the value of X is plotted in the vertical direction. This was done so that the function could be plotted against degrees. The cosine function is abbreviated in mathematical formulas as cos (a), which is the cosine of angle a, again expressed in degrees.

THE COSINE FUNCTION

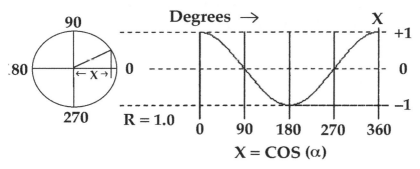

X = COS (α)

Figure 2-3. Developing the Cosine Function

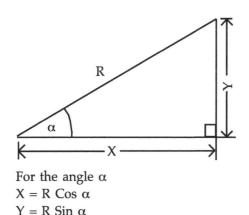

For the angle α
X = R Cos α
Y = R Sin α

Figure 2-4. Sine and Cosine Functions

Example of the Sine and Cosine Functions

Now that the first two basic trigonometric functions have been defined, it is useful to examine a practical use for these functions.

Example 2.1

If a 20-foot ladder were leaned against a building so that its base was 10 feet from the building, how far up the building would the ladder reach?

Referring to Figure 2-5 the length of the ladder is the radius, or 20 feet, and the X distance is 10 feet. Since the cosine of an angle is defined as X/R, it follows that:

Cos a = X/R   (2.3)
In this example:
Cos a = 10/20 = .5

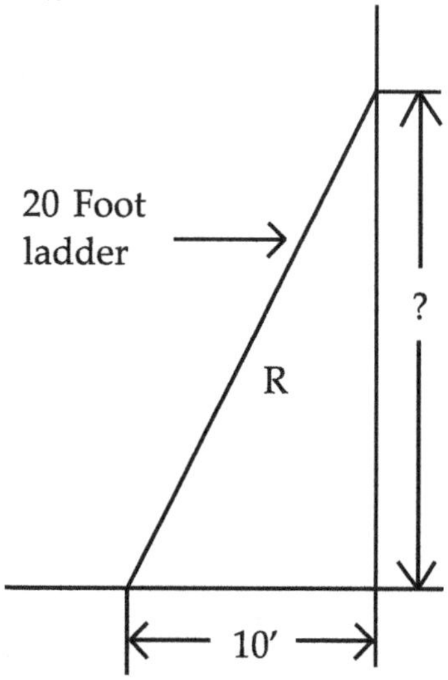

Figure 2-5. Using the Sine and Cosine Functions

Using the trigonometric table in Appendix C, the angle that has a cosine equal to .5000 is 60 degrees. Next look up the sin of 60 degrees, since the sine of an angle is defined as Y/R. The sine of 60 degrees is found to be .866025.

Sin (a) = Y/R   (2.4)

For this example:

Sin(60) = Y/R = .866025

By rearranging the equation,

Y = .866025 × 20 or Y = 17.325 feet.

DERIVING VIBRATION EQUATIONS

Referring to the Figure 2-6, the vertical displacement X at any time t can be expressed as:

$$X = Xo \; Sin \; \omega t \tag{2.5}$$

Where:
   Sin = the sine of the angle ω at time t
   $\omega = 2\pi f$ $\tag{2.6}$
   $f = 1/t$ [frequency in cycles per second]

Note: Cycles per second will be expressed in Hertz (Hz)
X_{max} occurs @ Sin ωt) = 1; Sin ωt =1 occurs at, $\pi/2, 3\pi/2, 5\pi/2, 7\pi/2$... $(2n-1) \; \pi/2$

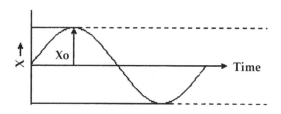

Figure 2-6. Basic Sine Wave

The velocity at any time t is expressed by:

$$V = dx/dt \tag{2.7}$$

And:
   $V = X_0 \; \omega Cos \; (\omega t)$ $\tag{2.8}$
   V_{max} occurs @ Cos $(\omega t) = 1$; Cos ωt =1 occurs at, $\pi, 2 \pi, 3 \pi,$ $4 \pi...n\pi$

Where:
   V_{max} = Maximum Velocity, See Figure 2-7.
   Cos = cosine of the angle ω @ time t

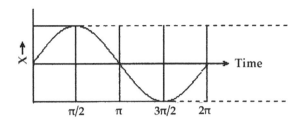

Maximum velocity @ π, 2π, 3π...nπ
Maximum acceleration @ π/2, 3π/2, 5π/2...(n–1)π/2

Figure 2-7. Maximum Velocity and Maximum Acceleration in a Sine Wave

If the capital letter F stands for the frequency in cycles per minute, then:

$$F = f * 60 \tag{2.8}$$

and:

$$f = F/60 \tag{2.9}$$

Letting the capital letter D stand for the peak-to-peak displacement in mils, then:

$$D = 2 \times X_o \tag{2.10}$$

and:

$$X_o = D/2 \text{ [D in mils]} \tag{2.11}$$

$$X_o = (.001 \ D)/2 \text{ [D in inches]} \tag{2.12}$$

It now follows that the relation between the maximum velocity and the peak-to-peak displacement can be expressed by the following equations:

$$V_{max} = X_o \ \omega$$

$$V = \frac{.001^*D}{2} \times \frac{2°F}{60} \text{ (inches per second)} \tag{2.13}$$

$$V = \frac{D*°F}{60,000} \tag{2.14}$$

$$D = \frac{60,000}{°F} \tag{2.15}$$

Acceleration is the second derivative of displacement with respect to time, which is expressed as:

$$A = dx^2/dy^2 = X_o \, \omega^2 \, Sin \, (\omega t) \tag{2.16}$$

A_{max} occurs @ $Sin (\omega \, t) = 1$; $Sin (\omega t) = 1$ @ $\pi/2, 3\pi/2, 5\pi/2,...$
$(2n-1) \, \pi/2$

Letting the capital letter "G" stand for acceleration in terms of G's, then:

$$G = A/[32.2 * 12] \text{ (inches per second per second)} \tag{2.17}$$

$$G = A/386.4 \tag{2.18}$$

$$G = [X_o \, \omega^2]/386.4 \tag{2.19}$$

$$X_o = D/2 = [.001 * D]/2 \tag{2.20}$$

$$\omega^2 = [2*\pi*f]^2 \tag{2.21}$$

Substituting from Equation (2.8)

$$\omega^2 = [(2\pi F/60)]^2 \tag{2.21}$$

Substituting from Equations (2.20) & (2.21)

$$G = \left(\frac{.001 * D}{2} \times \frac{(2 * °F)^2}{60^2} \right)/386.4 \tag{2.22}$$

$$G = .014190252 * DF^2/1,000,000 \tag{2.23}$$

$$G = .014190252 * D * [F/1000]^2 \tag{2.24}$$

$$D = [G/.014190252] * [1000/F]^2 \qquad (2.25)$$

$$D = [70.47090965 * G * [1000/F]^2 \qquad (2.26)$$

Now using the above relations to build the relation of acceleration to velocity:

$$V = X_o\, \omega \qquad (2.27)$$

And;

$$A = X_o\, \omega^2 \qquad (2.28)$$

And;

$$G = A/386.4 \qquad (2.29)$$

Thus:

$$G = X_o\, \omega2/386.4 \qquad (2.30)$$

$$G = X_o\, \omega^* \, \omega/386.4 \qquad (2.31)$$

And substituting from Equation (2.27):

$$G = V\, \omega/386.4 \qquad (2.32)$$

$$G = V^*(2\pi f)/386.4 \qquad (2.30)$$

And:

$$G = 2\pi F V/(60 * 386.4) \qquad (2.31)$$

Rearranging:

$$G = FV\pi/11{,}592 \qquad (2.32)$$

Or:

$$G = FV/3{,}689.848201 \qquad (2.33)$$

$$V = 3{,}689.848201 * G/F \qquad (2.34)$$

The use of these equations will transform the three forms of vibration characteristics to one another.

Example 2.2

A shaft has a peak-to-peak displacement reading of 3.4 mils @ 36.5 Hz. What is the velocity and what is the acceleration?

Using Equation 2.14:

$f = 36.5$ and $F = 36.5 \times 60$

$V = [\pi DF]/60{,}000 = [3.14159 \times 3.5 \times 36.5 \times 60]/60000$

$V = .401$ inches per second

Using Equation 2.24

$G = .014190252 * D * [F/1000]^2$

$\quad = .014190252 \times 3.5 \times [36.5 \times 60/1000]^2$

$G = .2382$

Using Equation (2.18)

$A = 386.4 * G = 92.04$ feet per second per second.

If the shaft weighed 23.6 pounds, it would have a mass of:

$M = W/g = 23.6/32.2 = .7329$

$F = MA = .7329 * 92.04 = 67.46$ pounds.

Thus the force generated by this vibration is almost three times the weight of the shaft. Using the vibration severity table at the end of Chapter 3, a velocity reading of .401 inches per second rates as rough, and some action should be taken.

Example 2.3

A centrifugal pump operating at 3,450 rpm shows a filter in vibration of .114 inches per second at operating speed. What is the peak-to-peak displacement? If the impeller and shaft weighs 40 pounds, what is the force produced?

Using Equation (2.15)

$D = 60000 \times .114/(3.14159 \times 3450) = .63$ mils peak-to-peak

The mass of the rotating element is $40/32.2 = 1.24$ pounds mass.

Using Equation (2.33)

$G = 3450 \times .114/3689.8484201 = .1066$

The force generated is:

$F = 386.4 \times G = 368.4 \times .1066 = 41.19$ pounds force.

Thus the force generated by this vibration is almost three times the weight of the shaft. Using the vibration severity table at the end of Chapter 3, a velocity reading of .401 inches per second rates as rough, and some action should be taken.

Example 2.3:

A centrifugal pump operating at 3,450 rpm shows a filter in vibration of .114 inches per second at operating speed. What is the peak-to-peak displacement? If the impeller and shaft weigh 40 pounds, what is the force produced?

Using Equation 2.15

D = 60000 × . 114/(3.14159 × 3450) = .63 mils peak-to-peak

The mass of the rotating element is 40/32.2 = 1.24 pounds mass.

Using Equation 2.33

G = 3450 x . 114/3689.8484201 = .1066

F = 386.4 × G = 368.4 × . 1066 = 41.19 pounds force.

Chapter 3

Resonance and Beat Frequencies

RESONANCE

To have a better grip on operating equipment, it is necessary to understand resonance and beat frequencies. All machines and machine components act like a spring-mass system. They vibrate due to some external force acting upon them. The vibration can either be free vibration or forced vibration. Figure 3-1 illustrates free vibration.

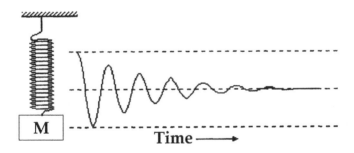

Figure 3-1. Free Vibration

Free vibration is simply an object vibrating at its natural or resonant frequency. In the figure above, the mass was displaced to the top of its motion and released. It vibrates at its natural frequency. Since no additional external force is added, the amplitude of the vibration decreases gradually with time.

One characteristic of a vibration occurring at a natural frequency is that very little force is required to maintain the vibration, as compared to the force required to maintain a vibration at a frequency other than the natural frequency.

If the applied force is exactly in phase with the natural frequency, the amplitude will increase dramatically. This is true even if the force is relatively small.

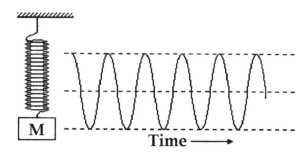

Figure 3-2. Vibrating at Natural Frequency

Two factors influence the natural or resonant frequency of an object, its mass and its stiffness. For a given object, adding mass lowers its natural frequency, and increasing the stiffness of an object increases its natural frequency.

Example 3.1

A 20-pound object vibrates at its natural frequency of 25 Hz. It is decided to add 5 pounds to the object to change its natural frequency. What will be the resulting new natural frequency?

The object originally has a mass of 20/32.2 or .62111 pounds. Its new mass will be 225/32.2 or .77639 pounds. The original frequency can be described by:

$$f = 1/(2\pi(m/k)^{.5} \tag{3.1}$$

Rearranging to solve for the constant k:

$$k = m \, (2\pi f)^2 \tag{3.2}$$

Since the original frequency was 25 Hz and the original mass was .62111 pounds, using Equation (3.2) yields k = 15,325 lb./ft. Now, using the constant k and the new mass, and substituting into

Equation (3.1) yields f =22.36 Hz. This is only a 10.6% change in frequency for a 40% increase in mass.

In most cases, adding mass to a component is not practicable and the component must be stiffened.

Example 3.2

In the system described in Example 3.1, the stiffness of the object is increased by welding a bracket in place. The bracket weighs 1 pound and increases the stiffness to 28,750 lb./ft. What is the resulting new natural frequency?

The new mass is 21/32.2 or .6522 pounds. Using the new mass and the new constant k, and substituting into Equation (3.1) yields f = 33.42 Hz..

This change in the stiffness resulted in a 33.68% change in the natural frequency of the object. This frequency change would be adequate to keep the object from being excited by an external force at 25 Hz.

Of course, the best way to solve the situation is to remove the source of the original vibration. Generally, the vibrating source is operating at a fixed speed and a significant reduction in its driving force cannot be accomplished. Vibration isolators or additional dampening can sometimes be added.

Most vibrations in machine elements are of the damped type as illustrated in Figure 3-3. Although most parts do not have dampers or shock absorbers attached to them, contact with other components of a machine has the effect of dampening the vibration.

DAMPED VIBRATION

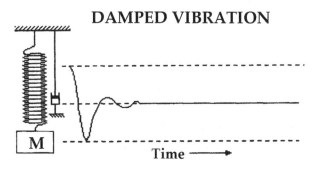

Figure 3-3. Damped Vibration

When a machine is in operation, there are normally many external forces that act upon the machine components. These repetitive forces are responsible for the vibration to appear as though it were free and not dampened. If one or more of these forces occurs at or near the natural frequency of one of the components, it will vibrate with an increased amplitude.

Resonant frequency in shafts or rotors is often referred to as a critical speed. Criticals are often experienced as a machine comes up to operating speed. This is often true for high-speed machines, which may pass through one or more criticals before reaching operating speed.

The first critical is simply the natural or resonant frequency of the vibrating element. There are also multiples of this natural frequency called the second harmonic, third harmonic, etc. These are also referred to as the second critical and third critical, etc.

Figure 3-4 shows the wave shape of a component vibrating at its natural frequency.

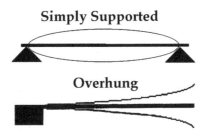

Simply Supported

Overhung

Figure 3-3. First Critical Vibration

Note the different shape of the wave for the simply supported component versus the cantilever or overhung component. Even without a vibration analyzer, feeling along the length of a component undergoing vibration at one of its critical frequencies, resonance can be detected. For the simply supported component, if the vibration is the highest at its midpoint, and decreases as either end is approached, the component may be vibrating at its natural frequency.

The cantilevered or overhung component will have its highest amplitude at its unsupported end. Simply marking the location of the highest perceived amplitude on a component with a felt marker can assist in determining the mode of the vibration.

If the frequency were to continue to increase above the natural or fundamental frequency, it would first decrease, and then increase as it approaches the second harmonic of the component.

At the second harmonic, the component again vibrates freely and the amplitude increases dramatically. The overall amplitude of the second harmonic is not as great as the amplitude of the first harmonic, but the frequency is twice as great.

Figure 3-5 depicts components undergoing second harmonic vibration.

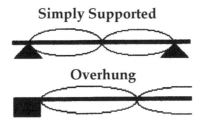

Figure 3-5. Second Harmonic Wave Forms

By examining Figure 3-5, it is easy to visualize what the third, fourth, and fifth harmonic would look like.

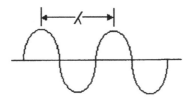

Figure 3-6. Wave Length

Letting (λ) represent the length of the wave, a system in resonance would have wavelengths of:

$$\lambda = 2L/n \tag{3.3}$$

Where:

L = length
n = 1,2,3,4...

The velocity of the wave can be expressed as:

$$v = f\lambda \tag{3.4}$$

Since the speed of compression waves is related to density and the elasticity of a material, the velocity may also be expressed as:

$$v = (E/p)^{.5} \tag{3.5}$$

Where:

E = Modulus of elasticity
p = Mass density

This equation states the velocity is equal to the square root of the modulus of elasticity divided by the mass density.

In Figure 3-7, a simply supported pipe is vibrating at its natural frequency. Note that at the two supports, the amplitude is zero. These are referred to as node points. In this case, the center has the greatest amplitude and is referred to as an anti-node.

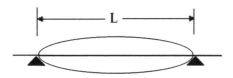

Figure 3-7. Pipe Vibrating at its Natural Frequency

Example 3.3

The pipe in Figure 3-7 was struck near its center and was found to vibrate at a frequency of 2 Hz. The distance between supports was measured to be 12 feet. What is the velocity of the wave? What frequencies should be avoided so that the pipe will not resonate at its natural frequency or one of the first three harmonics?

Step 1. Since the pipe is vibrating at its first harmonic, n = 1, and substituting into Equation (3.3), λ = 24 feet. From Equation (3.4), the velocity is found to be v = (24)(2) = 48 feet per second.

Step 2. By rearranging Equation (3.4) f = v/λ. The second harmonic λ_2 = L, and then the third harmonic λ_3 = 2L/3. Therefore, the second harmonic frequency is 96 Hz. and the third harmonic frequency is 144 Hz. Thus the three frequencies to be avoided are 48 ±10%, 96 ±10%, and 144 ±10% Hz.

If a frequency near these existed, consideration should be given to relocate the distance between pipe supports. Obviously, the distances 6 feet and 4 feet should be avoided.

Example 3.4

A steel shaft is simply supported with 4 feet between supports. The shaft is struck at its end and allowed to vibrate freely. What should be the recorded frequency?

Step 1. Since steel weighs 480 pounds per cubic foot, the mass of a cubic foot is (480/32.2 × 12))/1728 or .00071888 pounds mass per cubic inch. The modulus of elasticity for steel is approximately 30 × 10^6. Using Equation (3-5) the velocity is found to be 17,023.5 inches per second.

Step 2. Since the shaft is vibrating at its first critical or natural frequency, λ = 2L. Since L equals 4 feet, λ = 8 feet or 96 inches. Rearranging Equation (3-4), f = v/λ or the frequency anticipated is approximately 177.3 Hz.

For a machine, or machine component, vibrating at resonance or one of its harmonics, there are only three basic methods that can be employed to correct the vibration.

1. First, the external force that is exciting the component or machine can be eliminated or its frequency changed. This may not be practical in many cases since most equipment must operate at some given speed.

2. Second, by adding or removing mass. Usually, mass changes must be significant to alter the natural frequency sufficiently to eliminate the vibration.

3. This leaves altering the stiffness of the component. Care must be exercised when altering the stiffness of a component or machine, so that its strength is not also effected. Altering the location of support members will alter the effective length and alter the resonant frequency. This is usually a good choice for piping systems or other long, slender members. Use caution so as not to place a support at a location which will create other harmonics—i.e., 1/2, 1/3, etc.

It is always best to remove or at least decrease the source or driving force of one or more of the vibrations.

The easiest way to measure the natural frequency of an object is to first remove all sources of excitation. Next, the object is struck and allowed to vibrate freely. Using a vibration instrument, the frequency is recorded. The largest amplitude measured with a filter in reading will be the natural frequency.

BEAT FREQUENCIES

Beat frequencies are vibration frequencies caused by two or more vibrating sources that are vibrating at slightly different frequencies. The resultant beat frequency is the difference in the two sources.

Example 3.5

A machine is found to produce a vibration at a frequency of 1,275 Hz. A second machine nearby produces a vibration at 1,280 Hz. What is the resulting beat frequency?

Step 1. If at time zero, both machines were in phase, that is they were both at the maximum amplitude, the resulting beat would be at its maximum. After $1/10^{th}$ of a second, the first machine would have completed 127.5 cycles and be at a minimum, while the second machine would have completed 128 cycles and be at a maximum. If the two amplitudes were equal, the vibrations would be 180 degrees out of phase and thus cancel each other. At $1/5^{th}$ of a second, the first machine would have completed 255 cycles and the second machine 256 cycles. Both machines would now be at a maximum and the amplitudes would add together.

It can be seen that at time $1/10^{th}$, $3/10^{th}$, $5/10^{th}$ ... will produce minimums while $2/10^{th}$, $4/10^{th}$, $6/10^{th}$ ... will produce minimums. Thus the beat frequency is produced at 5 Hz. Note that this is the difference in the two original frequencies.

A component that is vibrating close to its natural frequency or one of its harmonics can exhibit beat frequencies as well. The beat frequency is the result of the two amplitudes coming in and out of phase with one another. When they are in phase, their amplitudes are added together; when they are out of phase, they subtract from one another.

The result is a beat frequency with an amplitude greater than either contributing source, and a frequency that is comprised of the multiple vibration amplitudes adding and subtracting from one another. The frequency of the wave form depends on amplitudes, frequency and phase angle of the contributing sources.

A simple beat frequency is shown in Figure 3-8, where two vibrating sources of different frequencies, amplitudes, and phase angles are contributing to the beat wave. Beat frequencies come into and out of phase as the contributing waves add and subtract from one another. Note that the composite wave is not a perfect sine wave as are the contributing sources.

The actual frequency of a beat is the sum of two or more vibrating sources adding and subtracting from one another as illustrated in Figure 3-8. They can be eliminated by either removing one or more of the sources of vibration, or changing the frequency of one or more of the sources.

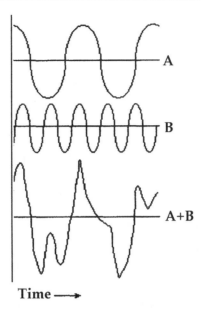

Time ⟶

Figure 3-8. Beat Frequency

If a part vibrates within 10% of its natural frequency or one of its harmonics, it may be subject to a beat frequency as well, even though there appears to be no second vibrating source.

In general, beat frequencies can be easily detected since they come in and out of phase, causing the vibration amplitude to increase and then appear to go away. Beat frequencies can excite the natural frequency of one another, or perhaps a third member will go into resonance.

Careful analysis will assist in determining the best way to eliminate problems with beat frequencies.

Chapter 4

Vibration
Due to Unbalance

INTRODUCTION

Unbalance is the most common source of vibration in rotating equipment. Vibration due to unbalance occurs at a frequency of 1 x rpm of the unbalanced element, and its amplitude is proportional to the amount of unbalance. Normally, the vibration signal is measured in both the vertical and horizontal planes. With a simply supported rotating element, that is a bearing at both ends of the shaft, the vibration due to unbalance will be in the radial plane with very little vibration sent in the axial direction. In the case of an over hung rotor, high axial vibrations may occur, and the amplitude in the axial plane may equal those measured radially. See Figure 4-9.

The force generated by unbalance can be expressed as:

$$F = 1.774 * (rpm / 1000)^2 * in\text{-}oz. \tag{4.1}$$

Where:

F = Force in pounds.
rpm = is the rpm of the unbalanced element.
$in\text{-}oz.$ = inch ounces of unbalance.

NOTE: An inch-ounce of unbalance is simply the amount of unbalance weight in ounces times the distance from the center of rotation, expressed in inches. Thus an inch-ounce can be a one-ounce weight at one inch from the center of rotation, or a tenth of an ounce at ten inches from the rotational center.

Rotating elements tend to rotate about their center of gravity, unless they are constrained. In a machine element, the center of rotation and the center of gravity should be the same. When they are not, vibration due to unbalance will result.

TYPES OF UNBALANCE

Unbalance can be of four basic types, static, couple, quasi-static and dynamic. Static unbalance is shown in Figure 4-1.

STATIC UNBALANCE

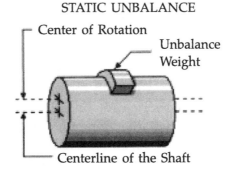

Figure 4-1. Static Unbalance

Static unbalance is defined as an unbalance where the center of rotation is displaced parallel to the geometric center of the rotating element. If the element were placed on knife-edges, it would rotate until the heavy spot was on the bottom.

With a statically unbalanced element rotating, the amplitude and phase of the vibration at both ends of the shaft would be the same. This static unbalance can be easily corrected by adding or removing the proper amount of weight as long as it is done in the proper plane. Figures 4-2 and 4-3 show the proper way to balance a pure static unbalance.

In Figure 4-2 a single weight equal to the unbalance is placed exactly in the same plane as the unbalance and exactly 180 degrees away. This brings the rotational center and the center of the shaft in-line.

CORRECTING STATIC UNBALANCE

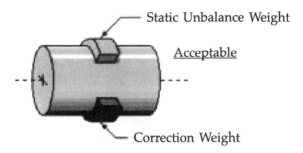

Figure 4-2. Correcting Static Unbalance

In Figure 4-3, two equal weights that equal the weight of the unbalance are placed equal distances from the unbalance and 180 degrees away. This also causes the rotational center to align with the center of the shaft.

The correction used in Figure 4-4 results in another type of unbalance, couple unbalance. Couple unbalance is defined as the condition of unbalance where the element's center of rotation passes through the element's geometric centerline at the element's center of gravity.

CORRECTING STATIC UNBALANCE

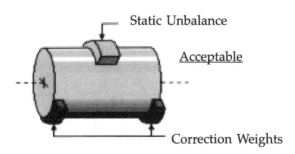

Figure 4-3. Correcting Static Unbalance with Two Weights

CORRECTING STATIC UNBALANCE

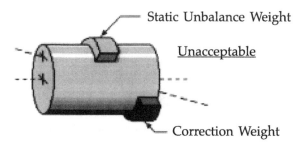

Figure 4-4. Creating a Couple Unbalance

Although the correction weight is equal to the unbalance and is placed 180 degrees away, it is not in the same plane. The piece is statically balanced, but when rotated the two weights act in different planes causing couple unbalance.

Couple unbalance is illustrated in Figure 4-5. Note that even though the shaft is statically balanced, the element would tend to wobble about its center when the shaft is rotated. Both ends of the shaft would vibrate with the same amplitude, but be 180 degrees out of phase.

COUPLE UNBALANCE

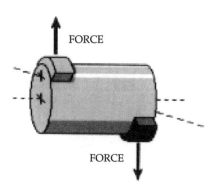

Figure 4-5. Forces in Couple Unbalance

Unlike static unbalance, couple unbalance cannot be detected by placing the element on knife-edges. Thus a couple unbalance can only be detected with the element rotating. The element is essentially statically balanced. A couple is basically two equal forces acting in opposite directions, and through two different planes. Again, unlike static unbalance, couple unbalance cannot be corrected in a single plane, but rather it requires corrections to be made in two or more planes. In only a few cases will either pure static or pure couple unbalance be detected in machinery. Most often it is a combination of both. These combinations are classified as dynamic and quasi-static unbalance. Figure 4-6 shows quasi-static unbalance. Note that the quasi-static unbalance is made up of a static unbalance and a Couple unbalance.

QUASI-STATIC UNBALANCE

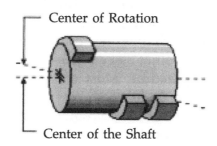

Figure 4-6. Quasi-Static Unbalance

Quasi-static unbalance is the condition of unbalance where the center of rotation intersects the element's geometric centerline, but not at its center of gravity. Quasi-static unbalance can be detected by the amplitudes of the vibration being very different at each end of the shaft, and being out of phase by approximately 180 degrees. Once again, this type of unbalance must be corrected in two or more planes.

The fourth type of unbalance is dynamic unbalance and is the most common type encountered in machinery. Dynamic unbalance is defined as unbalance where the axis of rotation does not coincide or touch the element's geometric centerline. Dynamic unbalance is depicted in Figure 4-7.

DYNAMIC UNBALANCE

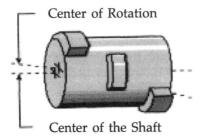

Center of Rotation

Center of the Shaft

Figure 4-7. Dynamic Unbalance

Dynamic unbalance most often exhibits different amplitudes of vibration at each end of the rotating element. In addition, most often it will exhibit phase angles that are neither in phase nor directly opposite from one another. This type of unbalance must be corrected in two or more planes.

Many of the conditions that cause unbalance in rotating equipment are the result of assembly in the field and/or maintenance and repair practices. Most of these conditions can be avoided, once they are understood. When a piece of equipment is to be worked on, it pays to have a complete understanding of how each part of the equipment was originally balanced, and in some more critical applications, how it was assembled.

Example 4.1

Assume a centrifugal compressor wheel was balanced on a dynamic balancing machine and was assembled with its keyway in the 12 o'clock position. Later when it was reassembled in the field, it was assembled with the keyway in the 6 o'clock position, causing the wheel to be assembled five hundredths of a thousandth of an inch from its true center (.00005"). The rotor weighed 20 pounds (320 ounces) and was turned by a Solar Saturn Gas Turbine at 23,200 rpm. What force was generated due to unbalance?

Step 1. This assembly error would result in the center gravity of the rotor being five hundredths of a thousandth of an inch

(.00005") from its balanced position to its field assembly position. Thus the inch-ounces of unbalance was 320 × .00005 = .016. Using formula (4.1) yields F = 1.774 × (23200/1000)² × .016 = 15.2 pounds of force. That is over 75% of its weight!

Although most of the equipment does not rotate at that high an rpm, much of the more common electric motor driven equipment does operate at 3,450 or 1,750 rpm.

Example 4.2

Steel weighs approximately 485 pounds per cubic foot or about 4.49 ounces per cubic inch. If an electric motor had a 3-inch shaft and rotated at 3,450 rpm, and had a 1/2" × 1/2" × 3" keyway that was not filled with key material, how much force would be generated?

3,450 RPM

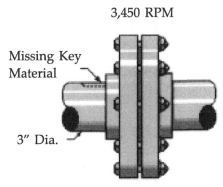

Missing Key
Material

3" Dia.

Figure 4-8. Example of Missing Key Material

Step 1. The missing key material is .75 cubic inches which weighs approximately 3.37 ounces. The center of mass for the key material would be 1-1/4 inches from the motor centerline. Thus 4.21 inch-ounces of unbalance would be present. Using the formula below to determine the force,

F = 1.774 × (3450/1000)² × 4.21 = 88.89 pounds of force.

Although most machines' bearings can withstand loads of up to eight times the normal static load, for a short period, it can be easily seen that the life of these parts will be greatly reduced.

SOURCES OF UNBALANCE

Most electric motor manufacturers standardize on balancing their rotating elements with the entire keyway filled with a half key. That is the shaft is totally filled. Coupling manufacturers practice the same procedures when balancing their couplings. When assembling these two elements, an L-shaped or T-shaped key must be used to maintain the original balance. Using the example above and viewing Figure 4-8 will illustrate the problems with not using the proper key during assembly.

As the person responsible for maintenance and repairs, it should be your goal to assure all possible sources of unbalance are inspected and corrected if necessary. The less vibration a machine has, the less stress it will be subjected to, and the longer it will survive. Equipment with a history of coupling, bearing, or seal failures should be inspected for sources of unbalance.

Figure 4-9 shows some of the more common sources of unbalance associated with coupling installation. Any of these conditions can lead to premature failure of bearings, couplings, or seals.

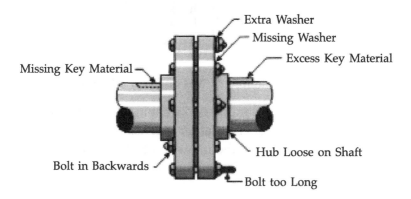

Figure 4-9. Sources of Unbalance

As in all maintenance activities, cleanliness is of extreme importance in maintaining the inherent balance of rotating machinery. A small particle of dirt and subsequent component failure.

When performing maintenance activities, care should be taken to clean all buildup of dirt and debris from rotating components, such as fan blades. Uneven buildup will result in an unbalance that will lead to early failure of bearings and seals.

Example 4.3

A 6-foot diameter nodular iron flywheel weighs 8,634 pounds, and is assembled onto an engine's crankshaft with bolts in a 12-inch circle. During assembly, a piece of debris .001-inch thick wedges between the flywheel and the crankshaft. The unit operates at 300 rpm. How much force is generated as a result of this misalignment of the flywheel?

A cross-sectional view of the flywheel is shown in Figure 4-10.

QUASI-STATIC UNBALANCE

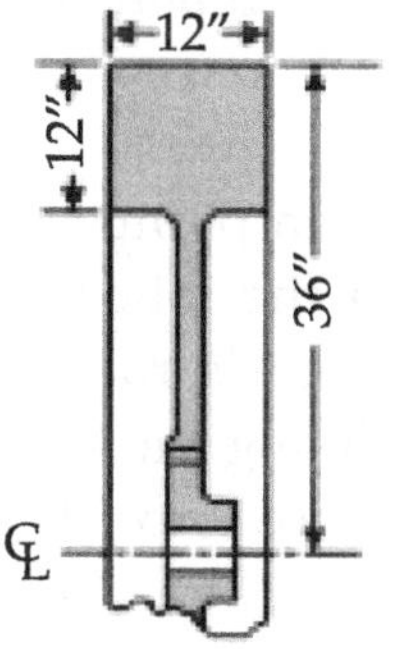

Figure 4-10. Cross-sectional View of the Flywheel

Step 1. The weight of the flywheel is displaced .001" from the center, thus the unbalance in inch-ounces is $U = 8634 \times 16 \times .001 = 138$ inch-ounces. The unbalance force is $F = 1.774 \times (300/1000)^2 \times 138 = 139.8$ pounds.

The above example illustrates that even low rpm machinery can have excessive forces due to unbalance. However, the higher the rpm the less the unbalance that can be tolerated.

Vibration due to unbalance occurs at a frequency equal to the rotational speed [1 × rpm] of the rotating element, and has an amplitude proportional to the amount of unbalance. The vibration will be largest in the radial direction, and only a single-phase mark will be present.

Equipment with overhung rotors may exhibit high axial vibrations due to unbalance. See Figure 4-11.

SIMPLY
SUPPORTED   OVERHUNG

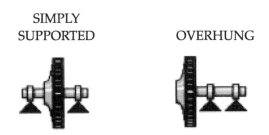

Figure 4-11. Types of Support for Rotating Elements

Example 4.4

An electric motor driven pump operates at 1,750 rpm and a velocity reading taken on the inboard motor bearing was .106 inches per second. The motor's rotor weighs 24 pounds. If the vibration was due to a missing piece of key material in the coupling, what was the unbalance force? If the keyway was 1/2" × 1/2," how long was the missing key if the motor shaft was 4 inches in diameter?

Step 1. The velocity reading needs to be converted into acceleration and then into the force. Using Equation (2.33), G = F*V/3,689.848201. Since the vibration was due to unbalance, it was occurring at 1 × rpm or at a frequency of 1,750 cycles per minute. Thus, F = 1,750 cpm. Now, G = 1750 * .106/3689.848201 = .0527, and A = 386.4 * G = 386.4 * .0527 = 19.42 feet per second per second.

Step 2. The mass of the rotor is W/g = 24/32.2 = .745 pounds mass. F = MA = .745* 19.42 = 14.47 pounds force. Now the force due to unbalance is expressed as F = 1.774 * (rpm/1000)^2 * in-oz. Here, the inch-ounces of unbalance is to be determined, so rearranging the equation, in.-oz. = F/1.774 * (rpm/1000) 2 in.-oz. = 14.47/(1750/1000)^2 = 4.72 inch-ounces of unbalance.

Step 3. The center of the missing key material would be at a radius of 1-3/4 inches, thus the missing material would weigh 16.54/1.75 or 2.7 ounces. Steel weighs approximately 4.49 ounces per cubic inch, therefore the missing material would be 2.7/4.49 or .6 cubic inches. Since the key is 1/2 × 1/2, the missing portion would be 1/2 × 1/4 or 1/8 ounce per inch of length. The length of the missing material is .6/.125 or 4.8 inches.

In the case of electric motors, measure the amplitude and then turn off the power. If the vibration disappears immediately, the problem is most likely electrical. However if the amplitude decreases with the decrease in speed of the motor, check for balance and/or misalignment.

There are a number of published vibration severity charts that give guidelines as to how much vibration is too much. One major pitfall of using these guidelines is the relationship of the weight of the rotating element to the total equipment mass and stiffness.

A small rotating element in a massive machine that is well anchored may not exhibit severe vibration when readings are taken on the machine frame. Some vibration analyzer companies provide shaft riders which allow the vibration probe to be placed against the rotating shaft to obtain a more accurate reading.

Regardless of the method employed, it is obvious that the smoother a part runs, the longer the life of the equipment. Many conditions of unbalance are best left to shop balancing rather than field corrections. However, there are many conditions of unbalance that can be recognized and corrected during routine maintenance. Cleanliness and attention to detail can avoid many of these situations.

Table 4-1. Vibration Severity

Classification	Velocity (in/sec.)
Very rough	Above .628
Rough	.314
Slightly rough	.157
Fair	.0785
Good	.0392
Very good	.0196
Smooth	.0098
Very smooth	.0049
Extremely smooth	Below .0098

Chapter 5

Field Balancing

INTRODUCTION

Unbalance is present to some extent in all rotating equipment, and has been found to be the most common source of vibration encountered. Field balancing of rotating equipment depends on identifying the type of unbalance and then applying the correct balancing procedure. Single plane balancing is *only* successful in conditions where the unbalance is pure **static**. Other forms of unbalance will require two or more planes of correction.

Dual plane balancing will be required for the majority of equipment. A method of dual plane balancing is discussed later in the chapter.

Typical of the types of equipment that can be balanced by the single plane method are fin-fans and cooling tower fans. For a part to be successfully field balanced, two conditions must be met. First, the rotating part must be out of balance, and second, the required changes in weight must be able to be made to the rotating part.

The following single plane method of field balancing assumes the use of a vibration analyzer that has a strobe light to identify the phase angle. Another single plane balancing method that only requires amplitude will be discussed later.

Balancing a piece of equipment with it mounted and operating in its normal position is called in-place or field balancing. For much equipment, in-place balancing has many advantages over shop balancing. These include less cost; elimination of possible damage due to disassembly, transportation and reassembling; and extensive down time. Some machines, due either to their high speed or to the class of tolerable vibrations, will require precision shop balancing.

Balancing in-place is a very straightforward process, and involves following only a few simple rules. Unbalance is defined as an unequal distribution of mass about a rotating center point. Unbalance is further defined by International Organization of Standards (IOS) as, "That condition which exists in a rotor when vibratory force or motion is imparted to its bearings as a result of centrifugal forces." Unbalance will result in the vibration of the rotating part and its supporting bearings and associated support members.

Balancing is the process of identifying the amount and location of a heavy spot on a rotating part, and then removing or adding the correct amount of weight at the correct location to cancel the centrifugal forces caused by the unbalance. The amount of unbalance that exists is determined by the amplitude of vibration of the rotating part. The location of the unbalance is determined with the use of a strobe light to identify a phase location with respect to any existing reference point on the rotating element.

SINGLE PLANE BALANCING

The strobe light is configured to trigger on the vibration signal, at its peak. This point corresponds to the location of the heavy spot at some time in the rotation cycle. The exact location of the phase reference point is of no consequence in the beginning.

Next, a trial weight is attached to the rotating member, and its location, amount and distance from the center of rotation are noted. By adding the trial weight, the location of the resultant heavy spot will be altered. Thus, the amplitude of the vibration and its phase location will have been altered.

When a weight is added to a perfectly balanced rotating element, it will vibrate at a frequency equal to the rotating speed of the element and have an amplitude proportional to the unbalanced weight. A reference mark for phase will appear to stand still at some location under a strobe light triggered by the vibration. To further illustrate this point, if a 4 ounce weight were added to a balanced rotor, and caused a 3.5 mils vibration at 60 degrees, doubling the weight to 8 ounces would increase the amplitude of the vibration to 7 mils, but the phase angle would remain at 60 degrees.

Further, if the 4-ounce weight had been moved 60 degrees counterclockwise, the amplitude would have remained the same, but the phase reference angle would have shifted 60 degrees clockwise to 120 degrees.

The three fundamentals of balancing are:

1. The amount of vibration is proportional to the amount of unbalance.

2. The reference mark (phase angle) shifts in a direction opposite to a shift in the heavy spot.

3. The angle the phase mark shifts is equal in degrees to the angle the heavy spot was shifted.

The unbalance in a rotating element at the start of a balancing process is referred to as the original unbalance, and thus the associated phase angle and amplitude readings are called the original readings. Polar graph paper is very useful in laying out balancing problems. However, by using a protractor and a convenient scale the same results can be achieved.

As an example, a rotor has a vibration amplitude of 3.5 mils at 60 degrees, and its original readings are shown in Figure 5-1. Note that the scale chosen is 1/2 mil per division, but any convenient scale can be used.

The radial lines that radiate from the center represent the angular position, and the concentric circles are spaced 1/2 mil apart. This makes it convenient to visualize the problem.

When a trial weight is added to the rotor, one of three things must happen:

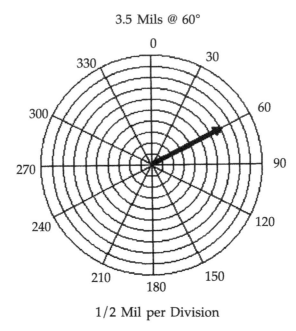

3.5 Mils @ 60°

1/2 Mil per Division

Figure 5-1. Graphical Solution with Original Vibration Plotted

1. The trial weight could be added exactly on the heavy spot. If
    so, the vibration amplitude would increase, but the reference
    mark would remain in the same location. To balance the ro-
    tor, the trial weight would simply be moved 180 degrees and
    adjusted in mass until the rotor was balanced.

2. The trial weight could be placed in exactly the correct posi-
    tion, 180 degrees away from the heavy spot. In this case, if
    the trial weight were smaller than the heavy spot, the vibra-
    tion amplitude would decrease, while the reference mark
    would remain fixed. Again, the mass of the trial weight
    would be altered until the rotor was balanced. However, if
    the trial weight were heavier than the heavy spot, the refer-
    ence mark would shift 180 degrees. The mass of the trial
    weight need only be reduced until a satisfactory balance was
    achieved.

3. Most often, the trial weight is placed neither on the heavy spot, nor directly opposite it. In this case the amplitude may go up or down, but there would be a definite change in the location of the reference mark. The angle and direction the trial weight must be moved, and the correct amount of the weight must be determined from a vector diagram.

Returning to our example, a trial weight of 5 ounces is added and a new set of readings is obtained. In this case, the new amplitude is 5 mils and the new reference angle is 120 degrees. This is shown in Figure 5-2.

In this figure, the original vector is labeled O and the run with the trial weight installed is labeled O + T.

Next, a vector is drawn from the head of the O vector to the head of the O + T vector, and is labeled the T vector. This vector represents the amount of vibration due to the trial weight alone. The length and direction of the T vector need to be determined to know where to properly place the balance weight. The length of the T vector can be measured using the same scale as selected for

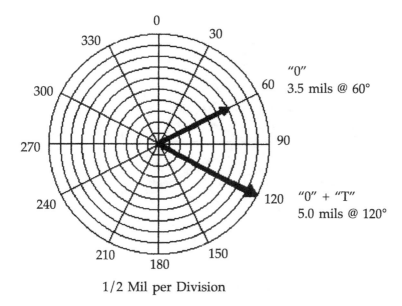

1/2 Mil per Division

Figure 5-2. Plotting the Trial Run Vector

the polar graph. The length of the T vector is the vibration amplitude in mils, as a result of the trial weight alone. Next, the angle between the O vector and the T vector is measured using a protractor.

Figure 5-3 shows the T vector correctly drawn.

With the T vector properly drawn, the length of the T vector is measured to be 4.4 mils. The angle α, the angle between the O vector and the T vector, is measured using a protractor, and found to be 79 degrees. With this information, we can analyze the unbalance problem.

First, we know that the trial weight caused 4.4 mils of vibration, and that a 5-ounce trial weight was added for the trial balance run. We can calculate the correct amount of weight to be added to the rotor with the following equation:

$$\text{Correction Weight} = \text{Trial Weight} \times \text{O/T} \qquad (5.1)$$

In this case, the correction weight = 5 × 3.5/4.4 = 3.977 ounces. The amount to be added would be rounded off to 4

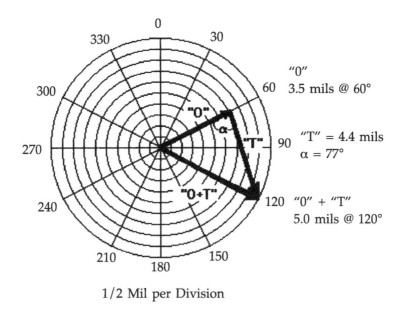

1/2 Mil per Division

Figure 5-3. Plotting and Measuring the T Vector

ounces. Next, the exact location for the correction weight must be determined. Again, we will use the information obtained in our trial run.

The object of the balancing process is to adjust the length of the T vector to equal the length of the O vector, and to adjust the direction of the T vector to be 180 degrees opposite of the O vector. The T vector can be thought of as a vector whose tail is always connected to the head of the O vector. Thus to make the T vector point exactly opposite the O vector, it must be shifted by the angle between the O vector and the T vector. This angle α was measured to be 79 degrees.

This is the angle that the trial weight must be moved. In Figure 5-3, the O + T vector was at 120 degrees, while the O vector had been at 60 degrees. Thus the reference mark shifted from 60 degrees to 120 degrees in a clockwise direction. Remember, the weight is moved opposite the reference shift direction, so the correction weight must be added 79 degrees counterclockwise from the position of the trial weight.

CAUTION: The correction weight is always shifted in an angle opposite in direction from the shift of the reference mark. The angle is measured from the location of the trial weight and not the reference mark.

REMEMBER: Remove the trial weight after adding the correction weight.

LAW OF SINE

The solution to the balancing problem in the example above can also be solved with the trigonometric functions sine and cosine. To accomplish this, the law of sine and the law of cos must be used. Refer to Figure 5-4.

Note in Figure 5-4 that the side **a** is opposite the angle α; the side **b** is opposite the angle β; and the side **c** is opposite the angle γ.

LAWS OF SIN & COS

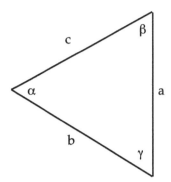

Figure 5-4. Relationships Between Angles and Sides of a Triangle

$\sin \alpha / \sin \beta = a/b$	(5.2)
$\sin \alpha / \sin \gamma = a/c$	(5.3)
$\sin \beta / \sin \alpha = b/a$	(5.4)
$\sin \beta / \sin \gamma = b/c$	(5.5)
$\sin \gamma / \sin \alpha = c/a$	(5.6)
$\sin \gamma / \sin \beta = c/b$	(5.7)

LAW OF COSINE

$a^2 = b^2 + c^2 - 2bc \cos \alpha$	(5.8)
$b^2 = a^2 + c^2 - 2ac \cos \beta$	(5.9)
$c^2 = a^2 + b^2 - 2ab \cos \gamma$	(5.10)

In our example, the angle between the O vector and the O + T vector is β, and the opposite side is the T vector. The angle between the O vector and the T vector is represented by α and the opposite side is the O + T vector. Finally, the angle between the O + T vector and the T vector is represented by γ and the opposite side, the O vector. This is illustrated in Figure 5-5.

We know that β = 60 degrees, but we don't know anything about the other two angles. Also, we know a = 5 mils and c = 3.5 mils. We need to know the angle α and the length b. The law of

LAWS OF SIN & COS

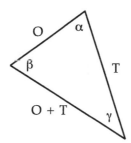

Figure 5-5. Solving the Balancing Triangle

cos is used first to determine the length of b (T vector). Since b^2 = $a^2 + c^2 - 2ac \cos \beta$ substituting yields $b^2 = 5^2 + 3.5^2 - 35 \cos (60)$, and $\cos(60) = .50000$ (from trigonometric tables). Thus $b^2 = 19.75$ or b = 4.44409.

Using the law of the sine the angle α is found by $\sin \alpha / \sin \beta = a/b$ or $\sin \alpha = \sin \beta \times (a/b)$. Substituting $\sin \alpha = \sin(60) (5/4.44409)$, and $\sin(60) = .866025$ (from trigonometric tables). Thus, $\sin \alpha = .866025 \times 1.125089 = .974356$ and $\alpha = 77$ degrees (from trigonometric tables).

Note that the answers obtained with the mathematical solutions are more accurate than answers obtained graphically. However, the two answers are so close that the resultant correction weight and its location will be about the same.

REVIEW

The basic steps required in the single plane balancing process:

1.   Operate the rotor at the balancing speed, and with the analyzer filter tuned to 1 × rpm, measure the amplitude of the vibration in mils, and using a strobe light, note the position of the phase reference mark. Record the amplitude and phase angle as the original run.

2. Stop the rotor and attach a trial weight to the rotor at a diameter where the correction weight can be added or removed. Record the amount and location of the trial weight.

3. Operate the rotor at balancing speed and again record the vibration amplitude and the phase angle of the reference mark. These data are the O + T run.

4. Using polar graph paper, construct both the O and the O + T vectors. Connect the head of the O vector to the head of the O + T vector and label the new vector T. **Note:** assure the T vector points from the O vector to the O + T vector.

5. Measure and record the length of the T vector. This is the vibration amplitude caused by the trial weight. Determine the amount of correction weight required by:

    Correction Weight = Trial Weight × O/T

6. Use a protractor to measure the angle between the O and the T vectors, and record its value. The correction weight will need to be placed this angle away from the trial weight location, in a direction opposite to the direction the reference mark shifted.

7. Be sure to remove the trial weight after securing the correction weight in place. Run the rotor again and record the amplitude of vibration and the reference mark location.

If additional balancing is required, this run will become the original data. Follow the steps 1 through 7 above until satisfactory results are achieved.

TRIAL WEIGHTS AND FORCE

Care should be exercised in the selection of the proper trial weight size. If it is too small, no change in phase or amplitude may be noted, and one trial run will have been wasted. If the trial weight is too large, it could cause damage to the machine, especially if it normally operates above the first critical speed. The subject of critical speed will be discussed later.

As a general rule, a trial weight should be selected that will cause a 30% change in amplitude or a 30-degree shift in the reference mark location. Changes of this magnitude will insure accurate results are obtained from your calculations. A common practice used in selecting the size of a trial weight, is selecting a mass that will produce a force equal to 10% of the rotor weight on the supporting bearings.

$$F = 1.774 \times (rpm/1000)^2 \times \text{inch-ounces} \qquad (5.11)$$

Example 5-1

A rotor turns at 1750 rpm and weighs 175 pounds. The rotor is simply supported by two bearings. Since each bearing supports 87.5 pounds, a force of 8.75 pounds will be sufficient to add the 10% force required for balancing. What trial weight is required?

Step 1. Using Formula 5.11, rearranging and substituting the supplied data, the unbalance is found to be inch-ounces = $8.75/[1.774 \times (1750/1000)^2]$ or inch-ounces = 1.61.

Step 2. If the trial weight were to be added at a radius of six inches, a weight of .268 ounces would be required. For practical purposes, a .25 or quarter ounce weight would be selected.

Be sure to attach trial weights securely to assure they do not fly off during the subsequent trial runs. There are numerous ways to add trial weights to a rotor, and a close inspection of the exact application will reveal the most appropriate method. If the rotor has a recessed lip, modeling clay can be used for trial weights. Flat washers, lead weights, hose clamps, reinforced tape, epoxy, and custom-made clamps have all been used as trial weights.

Trial weights and especially permanent correction weights can be added by soldering, brazing, and welding. Adding washers to existing assembly bolts or drilling and tapping new bolt holes may also be used. Occasionally there will be no location for trial weights. In this situation, a balancing ring may be installed to aid in the balancing process. A balancing ring is simply a flat metal ring that can be bolted to the rotating assembly, and has provisions to add trial weights.

Note: the balancing ring must be in perfect balance to prevent interference with the balancing process, unless it is to remain in place permanently. If the exact correction weight is found, material of equal weight may be removed 180 degrees opposite the correction weight location. Drilling, grinding or cutting can be used for this purpose.

Cautions

When balancing, assure the rotor operates at the same speed throughout the balancing process. A change in speed, load temperature and other operating conditions can alter the results. If a change occurs start the balancing process over.

By following these steps and cautions, field balancing presents few problems. However, if repeated attempts to balance a rotor fail in providing satisfactory results, and the analysis data point to unbalance as the problem, there may be other problems with the machine. Some of these problems that are frequently encountered are:

1. LOOSE MATERIAL - Machines such as blowers and fans may accumulate dirt or water in recesses, hollow blades and hollow shafts. This debris may take a new position each time the machine is stopped and restarted. Of course this will affect the balancing procedure.

2. ROTOR LOOSE ON ITS SHAFT - This problem is encountered on rotors that are pressed onto their shaft. If the interference fit is incorrect, the rotor may rotate slightly on the shaft during load changes due to the high torque during start up.

3. OPERATING IN RESONANCE - If the supporting structure or an element of the machine is resonant at or near the oper-

ating speed, balancing is usually very difficult. When a machine is in resonance, it is usually very sensitive to very small changes in the location or amount of any trial weight.

4.   EXCESSIVE CLEARANCE IN BEARINGS - Excessive looseness or clearance in bearings will sometimes cause the system to respond similarly to that experienced when operating in resonance.

EQUIVALENT WEIGHTS

There are times when the final correction weight must be placed at a different radius than the trial weight. When this is the case, a new correction weight mass must be calculated for the new radius. This is a straightforward process, since the balance correction is a product of both the mass and the radius (inch-ounce). All that is required is that the product of the radius times the weight remain the same.

$$Wt._1 \times R_1 = Wt._2 \times R_2 \tag{5.12}$$

Example 5-2

If a 2.4-ounce correction weight were added to a cooling tower fan at a radius of 54 inches, what weight would be required at a radius of 18 inches?

Step 1. By rearranging Equation (5.12) and substituting the supplied data, $Wt._2 = 2.4 \times 54/18 = 7.2$ ounces. Thus the 7.2-ounce weight at 18 inches produces the same amount of unbalance as the 2.4-ounce weight at 54 inches.

There are times when several weights can be combined into one, or times when a single weight must be split into equivalent weights to accommodate a particular piece of equipment. Once again, this task is very straightforward.

First let's examine the case of combining the weights. Figure 5-6 shows three weights being combined into one equivalent weight.

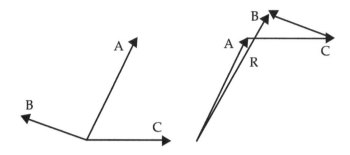

Figure 5-6. Vector Addition

As an illustration, assume 0 degrees is at the 3 o'clock position and that the degrees are laid off in the counterclockwise direction. Weight A is 11.5 ounces @ 63 degrees; weight B is 7.0 ounces @ 160 degrees; and weight C is 8.5 ounces @ 0 degrees.

When the three vectors that represent the weights are placed tail to head, and the resultant R drawn from the tail of the first vector to the head of the last vector, the measured length of the resultant is the equivalent weight. The angle formed by the resultant is the equivalent angle where the weight is to be applied. In this instance, the equivalent weight is 14.8 ounces @ 60 degrees.

Once again the same results could have been obtained using trigonometric functions. First, we can find all the vertical distances and add them together and then the horizontal components can be found and summed together.

$$Y = A \sin \alpha + B \sin \beta + C \sin \gamma + ... \qquad (5.13)$$

$$X = A \cos \alpha + B \cos \beta + C \cos \gamma + ... \qquad (5.14)$$

Therefore:

Y = 11.5 sin(63) + 7 sin(160) + 8.5 sin(0) or Y= 11.5 × .89100 + 7 × .342020 + 8.5 × 0.0 = 12.6407

and

X = 11.5 cos(63) + 7 × cos(160) + 8.5 × cos(0) or X = 11.5 × .45399 + 7 × (−.93969) + 8.5 × 1.000 = 7.14303

Since the values of X and Y form a right triangle:

$$a^2 + b^2 = c^2 \qquad (5.15)$$

By rearranging Equation (5.15) and substituting the supplied data,
$c = ((12.6407)^2 - (7.14303)^2)^{.5} = 14.585$

The length c is the resultant, or the required balance weight. In this case the weight is 14.6 ounces. Once again, the mathematical method is more accurate than the graphical method, but the answers are very close.

To find the angle, several methods could be employed, but we will use the arcsine (asin) function.

$$\text{Degrees} = \text{asin } (Y/R) \qquad (5.16)$$

Here, asin (Y/R) simply means the angle in degrees, whose sin = (Y/R). Dividing the value of Y by R and looking in the sin table for that value yields: asin $(.8669) = 60.07$ degrees.

Once again the mathematical solution is more accurate than the graphical method, but not enough to warrant its use in most cases. It is a good practice to work all problems using both methods to check your work.

The following illustration is provided to show that the planes onto which the weight can be resolved do not have to be at right angles to each other. Figure 5-7 represents a typical case of splitting the required correction weight into two equivalent weights.

The lines marked B, B', C and C' are not vectors, but lines representing planes. Note that vectors must have arrow heads indicating their direction of application.

To graphically resolve this problem, either the B plane or the C plane can be moved parallel, forming either the B' or the C' planes. These planes represent the direction the equivalent weights must act in, and therefore, must maintain the same relative angle.

Note that regardless of which plane was moved, it was

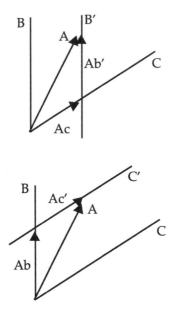

Figure 5-7. Splitting the Correction Weight

placed at the head of the A vector. Starting at the tail of the A vector, a vector is constructed by drawing an arrow head on the appropriate plane where it is intersected by the second plane where an equivalent weight is to be placed.

As with the other graphical problems, a mathematical solution is also possible, and will provide a more accurate answer. To follow through with this illustration, assume that vector A represented a 10-ounce weight, and acted at an angle of 75 degrees. Further, assume that the B plane is @ 90 degrees and the C plane is @ 30 degrees. Again, zero degrees is assumed at the 3 o'clock position and the degrees are laid off in a counterclockwise direction.

First, the A vector needs to be resolved into its X and Y components. Recalling $Y = A \sin \alpha$, and $X = A \cos \alpha$ or $Y = 10 \sin (75) = 9.65925$, and $X = 10 \cos (75) = 2.58819$. Now the equivalent values for each of the two planes B and C can be found. Since $Y = B \sin \beta + C \sin \gamma$ and $X = B \cos \beta + C \cos \gamma$, substituting the data yields $9.65925 = B \sin (90) + C \sin (30)$ and $2.58819 = B \cos (90) + C \cos$

(30). Solving the trigonometric functions yields 9.65925 = B + .5 C, and 2.58819 = .866025 C.

Now the two equations can be simultaneously solved by substituting the value of C from the first equation into the second equation. This yields C = 2.98858 and B = 9.65925 − (.5 × 2.98858) = 8.16495.

Thus by placing a 3-ounce weight at 30 degrees and an 8-ounce weight at 90 degrees will produce the same result as the 10-ounce weight at 75 degrees.

Note that although the B plane was @ 90 degrees, allowing a simple solution for C, the same results can be achieved regardless of the angles involved. There will always be two equations with two unknowns; therefore, a solution is easily achieved.

TWO PLANE BALANCING

The majority of unbalance situations are not pure static in nature; thus, utilization of single plane balancing is ineffective. In general, the unbalance must be either compensated for in the same plane in which it occurs, or resolved into two or more corrective planes to prevent the formation of couple forces.

In Figure 5-8 a correction weight equal to the unbalance is added in the wrong plane. The result will be a couple.

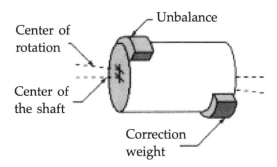

Figure 5-8. Locating the Correction Weight in a Different Plane than the Unbalance

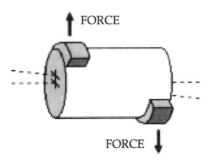

Figure 5-9. A Couple Force

The two forces are acting 180 degrees out of phase and in different planes, causing the shaft to wobble about its centerline. Note that the correction weight could also be a second source of unbalance and that it would not necessarily have to be directly opposite the unbalance nor equal in weight.

To prevent the formation of a couple force, the correction must act in the same plane as the unbalance, or resolve to the same plane. That is, the sum of the forces and the sum of the moments must equal zero. A moment is a force times the distance from a reference point.

In Figure 5-10 it can be seen that splitting the correction weights into two equal parts and spacing them equal distances from the unbalance is an acceptable solution to an unbalance. However, the number and location of the correction weights must always meet two basic criteria.

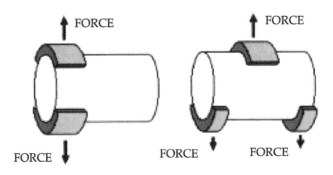

Figure 5-10. Two Acceptable Solutions to an Unbalance

1. The sum of the forces from the correction weight(s) and the unbalance, must equal zero.

2. The sum of the moments about the unbalance must equal zero.

Example 5-3

A rotating element has a 10-pound unbalance. A 5-pound weight is placed 5 feet to the right of the unbalance, 180 degrees opposite it. Two additional weights are added to the left of the unbalance, also 180 degrees opposite the unbalance, a 2-pound weight at 8 feet and a 3-pound weight at 3 feet. Show that the corrections made will balance the shaft.

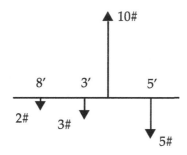

Figure 5-11. Example 5-3

Step 1. The two basic criteria must be met for the shaft to be in balance. Since the correction weights are all placed 180 degrees opposite the unbalance, the sum of their weights must equal the unbalance to satisfy the first criteria. 2 + 3 + 5 = 10, thus the amount of added weight is correct.

Step 2. The second criteria states the sum of the moments must equal zero. Selecting the point of unbalance as the reference point, the unbalance has no moment, since the distance is zero. Assuming measurements to the right of the reference are positive and measurements to the left are negative, the moments are 2 × –8 = –16, 3 × –3 = –9, and 5 × 5 = 25. Adding these moments yields – 16 – 9 + 25 = 0, and thus the second criteria is also met.

Referring to Figure 5-12, the unbalance on the shaft is closer to the left end bearing. Vibration readings taken at the two bearing locations will differ. Attempting to correct the unbalance at point A will not resolve the problem. Corrections will need to be made at both bearing locations.

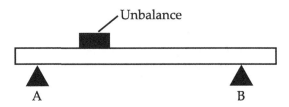

Figure 5-12. Unbalance on a Shaft

By observing Figure 5-12, it can also be seen that adding correction weights to either end will set up couple forces, and thus alter the vibration at the opposite end. Trying to correct at the second end will have the same effect on the first end and so on. This is referred to as cross-effect. Due to this cross-effect, each end cannot simply be treated as a single plane problem without having to make numerous runs and corrections. It is far better to determine the cross-effect and take its effects into account.

After it has been determined that a machine element requires balancing, and that a two plane approach is required, the machine should be shutdown and a convenient reference mark placed so that phase angles can be recorded. The end of a shaft is the best location, since it is easily observed during operation. In addition, shaft ends often have keyways or other distinctive features that can be identified for phase locations.

Some machines may require balancing in more than two planes to eliminate the forces of unbalance. This may be true of shafts that have multiple rotors stacked on them, where the unbalance may be in several planes. Although the two plane balancing method can satisfy the two basic criteria, if the rotor is flexible vibration still may occur.

A flexible rotor is considered as a rotor that operates within 70% or more of its resonant frequency or one of its harmonics.

These rotating elements will require balancing by a multi-plane method. This type of balancing is best left to shop balancing and not attempted in the field.

Multi-plane balancing is beyond the scope of this book and is not discussed further.

Since this balancing technique requires a bit more accuracy than the single plane method, it is recommended that a polar graph be attached to the machine by cutting out the center and placing it over the end of the shaft. This will assist in determining the exact phase angles during the balancing process.

Figure 5-13 shows a polar graph attached to a machine end, and its shaft end marked with an arrow. Felt-tip markers are useful in marking phase reference locations.

To start the balancing process, the vibration analyzer is tuned to the operating speed and a strobe light is attached to measure the reference phase angles. The use of two pickups is recommended to avoid having to move from one end to the other for each set of measurements.

An initial run is made and the amplitude and phase angles are measured and recorded for both the near end and the far end

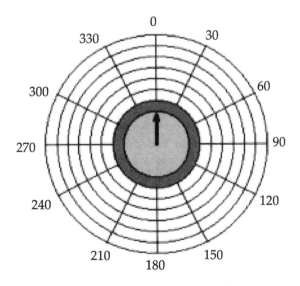

Figure 5-13. Using Polar Graph Paper for Phase Reference

of the machine. The machine is shutdown and a trial weight is added to the near end. Record the weight and the angle from the reference mark to the location of the trial weight.

Re-start the machine and measure and record the new amplitude and phase location for the near and far ends. Stop the machine and remove the trial weight. Place a trial weight at the far end, recording the weight and location. The location must be referenced to the near end and recorded as degrees clockwise from the reference point.

Re-start the machine and again measure and record the near- and far-end amplitudes and phase angles.

The remainder of the process of determining the amount and location of the correction weights is best explained through an example. Like the single plane method, the recorded data will be transferred to polar graph paper and additional information measured and calculated.

Example 5-4

A machine had the following data recorded: What are the recommended corrections?

Original readings:	**(N0)**	Near end: 7.1 mils @ 60 degrees.
	(F0)	Far end: 5.4 mils @ 215 degrees.
First trial run:	**(N1)**	Near end: 4.9 mils @ 120 degrees.
	(F1)	Far end: 3.7 mils @ 230 degrees.
	(T1)	Trial weight: 8 ounces @ 270 degrees
Second trial run:	**(N2)**	Near end: 5.1 mils @ 35 degrees.
	(F2)	Far end: 8.6 mils @ 160 degrees.
	(T2)	Trial weight: 10 ounces @ 180 degrees

What are the recommended corrections?

For this example, the following conventions will be used.

1. The convention N0 > N1 will represent a vector whose length is from N0 to N1 and the > sign denotes the direction in

which it acts. In this case, the vector is drawn from the tip of the N0 vector to the tip of the N1 vector.

Since vectors have two basic components, their magnitude and direction, the convention for a vector named A—calculated from two other vector quantities, X and Y—will be A = (X/Y) units @ (X + Y) degrees. In this case the desired length is the length of X divided by Y and the desired angle is X plus Y.

What is important is that the symbol for a vector will represent its length and the underlined symbol will represent the angle of that vector.

Step 1. Plot the six readings on polar graph paper. Figure 5-14 illustrates the plotted data. In this example, each circle represents 1 mil.

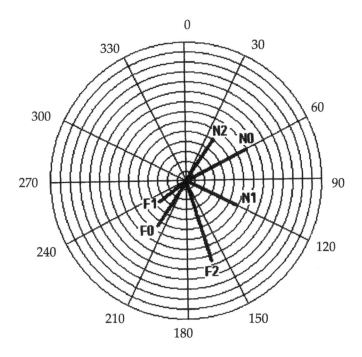

Figure 5-14. Example 5-4 Plotted Data

Step 2. Determine the length and direction of the vector from N0 > N1.

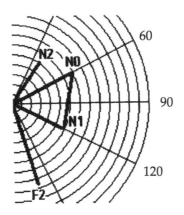

Figure 5-15. Example 5-4 Finding the Vector N0 to N1

Note the symbol > is used to denote the direction of the vector. Using the proper scale, the measured length of the vector from N0 > N1 is 7.6 mils. Drawing a line parallel to the N0 > N1 vector so that is passes through the center of the polar graph allows the angle to be measured. The angle is found to be 198 degrees.

The length of the N0 > N1 vector could have been calculated by breaking the N0 and N1 vectors into their X and Y components, taking their difference, and using the relation $r^2 = x^2 + y^2$. The angle can be found using the law of sines.

Step 3. The next step involves determining the length and direction of the F0 > F2; F0 > F1 and the N0 > N2 vectors. These can be plotted on the graph and measured in the same manner as the N0 > N1 vector.

Again, measuring the length and angles of these vectors yields:

<blockquote>
N0 > N2:   3.3 mils @ 281 degrees.
F0 > F2:   7.1 mils @ 121 degrees.
F0 > F1:   2.1 mils @ 7 degrees.
</blockquote>

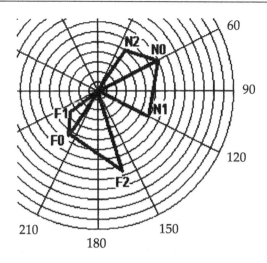

Figure 5-16. Example 5-4 Vector Construction

To further illustrate how these results can be achieved mathematically, the values for the F0 > F2 vector are calculated below. First, the original vectors F0 and F2 are broken into their X and Y values:

$$X = 8.6 \sin(160) - 5.4 \sin(215) = 6.0386$$

$$Y = 8.6 \cos(160) - 5.4 \cos(215) = -3.6579$$

$$r^2 = (6.0386)\, 2 + (-3.6579)^2 \text{ or } r = 7.1$$

Now that the length is known, the angle needs to be determined.

Referring to Figure 5-17, X, Y and r are known and it is further known that the angle c is a right angle. Using the law of sines, sin b = X/r sin c, or since c is 90 degrees, sin b = X/r. It follows that the angle b expressed in degrees equals arcsin(X/r). The angle b is found to be 58.26 degrees. From Figure 5-17, it is noted the angle e is equal to 180 – b or about 121 degrees.

In determining the angle, it is always useful to sketch the three vectors to determine an approximate angle to assure the results are correct.

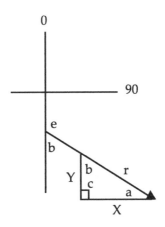

Figure 5-17. Determining the Vector Angle

Example (5.4) (*Cont'd*)

Step 4. Calculate the following additional vectors from the measured values. First, the measured values will be assigned new names.

A = N0 > N1   =   6.3 mils @ 198 degrees.
Aa = F0 > F1   =   2.1 mils @ 7 degrees.
B = F0 > F2   =   7.1 mils @ 121 degrees.
Bb = N0 > N2   =   3.3 mils @ 281 degrees.

The following calculations are now performed:

a = (Aa/A) mils @ (Aa – A) degrees = (2.1/6.3) @ (7 – 198) degrees.

a = .3 mils @ 171 degrees.

b = (Bb/B) mils @ (Bb – B) degrees = (3.3/7.1) @ (281 – 121) degrees.

b = .5 mils @ 160 degrees.

Na = (N0 × a) mils @ (N0 + a) degrees = (7.1 × .3) mils @ (60 + 171) degrees.

Na = 2.1 mils @ 231 degrees.

Fb = (b × F0) mils @ (b + F0) degrees = (.5 × 5.4) mils @ (160 + 215) degrees.

Fb = 2.7 mils @ 15 degrees.

Note: The sum of the degrees for Fb was greater than 360, so the result had 360 subtracted from it, 372 − 360 = 12. At times, the results can provide a negative degree. This is corrected by adding 360. Assume a reading of −45 degrees was obtained. The final answer would be −45 + 360 = 215 degrees.

Step 5. Once again, some vectors will need to be constructed. Starting with a new polar graph, construct the following vectors: Fb, N0, Na, F0.

Now the vectors C = N0 > Fb and D = F0 > Na can be plotted and measured for their length and associated angles.

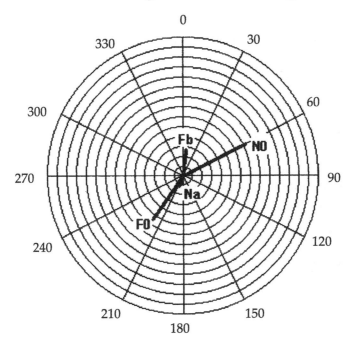

Figure 5-18. Example (5.4) Data Plot

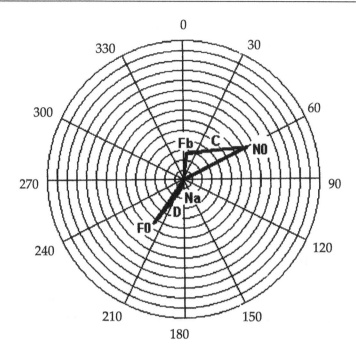

Figure 5-19. Example (5.4) Determining the C and D Vectors

Measuring the length and angles of these vectors yields:

C = 5.6 mils @ 258 degrees.
D = 3.2 mils @ 24 degrees.

Step 6. Again additional values need to be calculated.
ab = (a × b) units @ (<u>a</u> + <u>b</u>) degrees = (.3 × .5) units @ (171 + 160)
ab = .2 units @ 331 degrees.

Note: The ab vector is expressed in units not in mils. This will become clear further into this example.

Step 7. The next step is to use a new polar graph and plot a unity vector UV. The unity vector will be one unit in length at zero (0) degrees. This new plot should use a different scale from the other plots, allowing the unity vector to be as large as practical. For this example, the unity vector will be constructed at a scale of 10 to 1, that is 10 divisions on the polar plot will equal 1 unit.

Figure 5-20 illustrates the plot of the unity vector at a scale of 10 to 1.

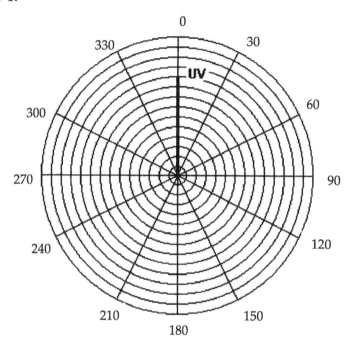

Figure 5-20. Example (5.4) Plotting the Unit Vector

With the unit vector plotted, the next step is to plot the ab vector on this new plot using the same scale as used to plot the unit vector. Next, construct the vector E = ab > UV and measure its length and angle.

E = .9 units @ 5 degrees.

Step 8. Now the final calculations may be performed.

Ac = (C/E) units @ (<u>C</u> – <u>E</u>) degrees = (5.7/.9) units @ (261
       – 5) degrees.

Ac = 6.4 units @ 253 degrees

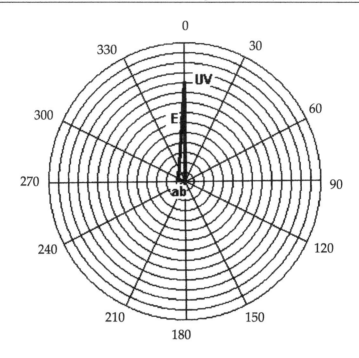

Figure 5-21. Example (5.4) Plotting the E Vector

Bd   = (D/E) units @ (D – E) degrees = (4.4/.9) units @ (36 - 5)

Bd   = 3.7 units @ 19 degrees

 c   = (Ac/A) units @ (Ac – A) degrees = (6.4/6.3) units @ (253 - 198) degrees

 c   = 1 units @ 56 degrees

 d   = (Bd/B) units @ (Bd – B) degrees = (3.7/7.1) units @ (19 - 121) degrees.

 d   = .5 units b@ 258 degrees

CWN = (T1 × c) ounces @ (T1 – c) degrees = (8 × 1) ounces @ (270 - 56)

CWN = 8 ounces @ 214 degrees

CWF = (T2 × d) ounces @ (T2 – d) degrees = (10 × .5) ounces @ (180 - 258) degrees

CWF = 5 ounces @ 282 degrees.

Note in the above calculations, the least significant digits were set as $1/10^{th}$. By calculating using least significant digits to $1/1000^{th}$, the answers are slightly altered. In this case, the solution would be:

CWN = 8.2 ounces @ 214 degrees.
CWF = 5.2 ounces @ 282 degrees.

Assuming the machine in example (5.4) operated at 1,750 rpm and the trial weights were placed at a radius of 16 inches from the center, the difference would cause an unbalanced force of: $F = 1.77 \times (1750/1000)^2 \times .2 \times 16$ or 17.34 pounds of force. Depending on the weight of the rotor and the type of machine, this may be significant.

WHICH METHOD

As a general guide, rotors with a W/D ratio of .5 or less may be balanced using the single plane method if they operate at less than 1,000 rpm. Above 1,000 RMP the two plane method should be used. Rotors with a W/D ratio above .5 that operate below 150 rpm may use the single plane method of balancing. Those above 150 rpm should use the two plane method.

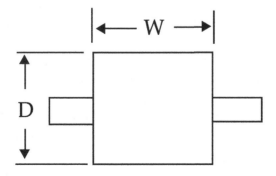

Figure 5-22. W/D Ratio for Rotors

OVERHUNG ROTORS

Overhung rotors are often found in fans or blowers that have a W/D ratio of less than .5, and most often operate at speeds of less than 1,000 rpm. The single plane method of balancing can often be employed to correct unbalance conditions for this type of rotor.

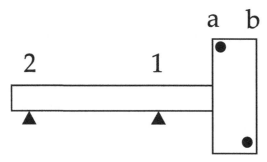

Figure 5-23. An Overhung Rotor

Referring to Figure 5-23, the vibration readings are taken at bearing 1 and the corrections are made in plane a. Once the corrections have been made, the vibration is measured at bearing 2. If the vibration levels are unacceptable, corrections must be made in plane b using the vibration data from bearing 2.

Adding a trial weight in the b plane will upset the balancing effort done in the a plane. To compensate for this, an equal but opposite weight is added in the a plane, to establish a couple force. This second trial weight is used to preserve the static balance in plane a.

Placement of the second trial weight in the a plane, must be exactly 180 degrees opposite the trial weight placed in the b plane and produce the same unbalance. That is, it is of the same weight and placed at the same radius, or of different weight but placed at a radius that creates the same inch-ounces of unbalance.

Now the single plane method is performed for the b plane using the data from bearing 2. Once the correction is made, be sure to remove the trial weight from the a plane, and then recheck its balance using data from bearing 1. These steps are repeated as often as required to achieve an acceptable balance of the rotor.

NOTES ON BALANCING

First and foremost, assure the machine is truly suffering from unbalance prior to attempting corrections to balance. Observe the machine for obvious causes of unbalance, such as broken or missing cooling vanes on electric motors, loose impellers, missing key material, excessive key material, different length bolts, etc. On fans, blowers and compressors, check for buildup of debris which can cause unbalance and clean first. Scale and rust buildup can also alter the balance of this type machine.

Many large cooling tower fans and fin-fan coolers have hollow blades. Usually, the tip of the blade has a small hole drilled to its center to allow any trapped moisture to escape. Assure these holes are open and free to flow any moisture from the blade. Internal corrosion and scale can also be a source of unbalance.

On fans with adjustable blades, assure they are all of the same length and are set to the same pitch. Another source on vibration that may appear to be unbalance is aerodynamic force generated as blades pass over support members.

Assure the major component of the vibration is occurring at 1 × rpm.

Chapter 6

A Single Plane Balancing Technique

INTRODUCTION

There is a simple balancing technique that is very useful in balancing cooling tower fans and fin fan coolers. This is the four-circle method and will be discussed in some detail. The basic advantage of this method is that it does not require a vibration analyzer that has a strobe light, or other methods of measuring phase angle. This method requires only amplitude measurements.

The main disadvantage of this method is that it requires starting and stopping the fan several times. However, it does work very well on these relatively slow-speed rotating elements.

GRAPHICAL SOLUTION

First, the amplitude of the vibration at 1 × rpm is measured and recorded. A circle with a radius equal to the measured amplitude is then drawn. Note that any convenient scale can be used for laying out the graphics, i.e., 5 mils equals 1/4 inch.

As an example, a cooling tower fan has three blades and operates at 330 rpm. An initial vibration was taken and the amplitude of vibration in the radial direction was recorded as 46 mils at 330 cpm. The vibration pickup was attached to the gear box and will remain attached at the same point throughout the balancing process.

The first step is to lay out a circle whose radius is equal to the amplitude of the original vibration. Remember, any convenient scale can be used, but be sure to allow for the additional circles which will be drawn with their centers located around the original circle's circumference. As a minimum, allow 4 times the original amplitude to be the minimum layout. That is, if the original amplitude was 46 mils, a minimum of 4 × 46 or 184 divisions should be available. Since the selected graph paper has 100 divisions, a scale of 5 to 1 was chosen. Thus, the required minimum would be (46/5)*4 or 36.8 divisions which will fit on the selected graph paper. **Note:** a scale of 2 to 1 could be used in this example. Using a scale of 5 mils per division, the circle would have a radius of 46/5 or 9.2 divisions.

A circle with a radius of 9.2 divisions was laid out to represent the original vibration amplitude, as shown in Figure 6-1.

Next, the fan is stopped, locked out, and the blades chained so that they can be numbered in a counterclockwise direction. Any blade can be designated as blade #1. The number of blades on the fan will determine the angle between blades. The angle between blades is simply 360/number of blades. The location of the fan blades is then laid out on the graph.

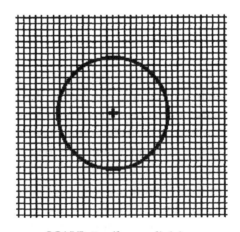

SCALE: 5 mils per division

Figure 6-1. Laying Out the Original Vibration Amplitude

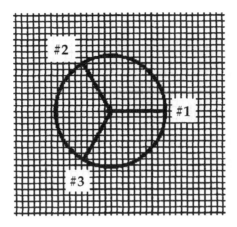

SCALE: 5 mils per division

Figure 6-2. Locating the Blades on the Graph

At 3 o'clock, draw a line from the original circle's center to its circumference, and label it blade #1. The angle between blades is measured from the blade #1 line to determine the location of blade #2. The same process is used to locate blade #3. Note that only the first three blades need be located. The angle between blades is calculated by:

$$e = 360/N_b \qquad\qquad (6.1)$$

Where:

N_b is the total number of blades on the fan.

A suitably sized trial weight is selected and secured to blade #1. The fan is operated and the amplitude of the vibration at operating speed is recorded. The amplitude of this vibration is drawn as a circle with a radius equal to the amplitude with its center at the point where the blade #1 line intersects the circumference of the original circle. In this example, the new amplitude was 31 mils peak-to-peak.

The fan is stopped and the trial weight is removed from blade #1 and placed on blade #2 at the same distance from the

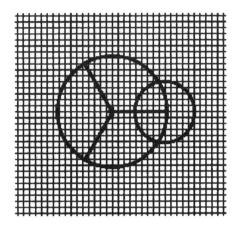

SCALE: 5 mils per division

Figure 6-3. Adding Trial #1 Vibration to the Graph

center as used on blade #1. Again, the fan is operated and the amplitude is recorded. In this example 49 mils peak-to-peak was recorded.

This amplitude is again drawn as a circle with its radius equal to the amplitude and its center where the blade #2 line intersects the original circle's circumference.

The same process is repeated for blade three, and the third trial-run circle is constructed. The graph should now consist of the original amplitude circle, three lines representing the first three blade locations, and three additional circles representing the amplitude of the three trial runs. In this example Trial Run #3 had a peak-to-peak amplitude of 71 mils.

Note that although part of the circles representing Trial Runs #2 and #3 did not fit on the graph, only the portion of the circle that intercepts the Trial Run #1 circle is required.

Once the basic graph has been laid out, the T vector can be drawn. The T vector is drawn from the original circle's center to the point where all three trial run circles intersect each other. It should be noted that all three circles may not intersect each other at exactly the same point, due to the error in amplitude measurements, or due to errors in construction of the graph. However, the

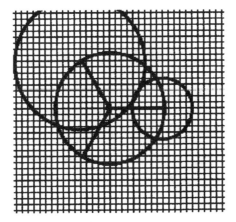

SCALE: 5 mils per division

Figure 6-4. Adding Trial Run #2 to the Graph

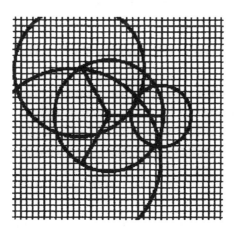

SCALE: 5 mils per division

Figure 6-5. Adding Trial Run #3 to the Graph

point can easily be estimated. Figure 6-6 depicts the correctly drawn T vector.

Due to scale factors and the thickness of the pencil, often the three circles don't all meet at the exact same point. Estimate the point where they should have met and calculate the T vector.

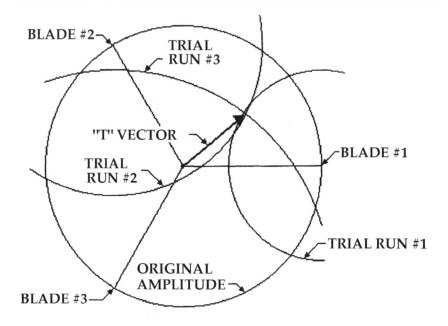

Figure 6-6. Drawing the T Vector

Figure 6-7 shows how to estimate the point where the three trial-run circles should have intersected each other.

Points A, B and C represent the intersection of circles 1 and 2, circles 2 and 3, and circles 3 and 1 respectively. To determine the exact location of the head of the T vector, three lines are drawn connecting the three circles' intersecting points, line A-A, line B-B, and line C-C.

These three lines will intersect at a common point. This point is the exact location of the head of the T vector.

In the case where the circles fail to overlap, draw a line from the center of circle 1 to the centers of circles 2 and 3. Next draw a line from the center of circle 2 to the center of circle 3. Divide each of these lines in half and draw a line from that point to the center of the circle opposite that line. Where these three lines intersect is the location of the T vector head.

The T vector is drawn from the center of the original amplitude circle to this point. The length of the T vector is now mea-

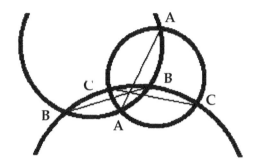

Figure 6-7. Estimating the Intersection Point

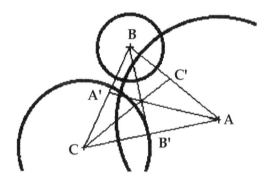

Figure 6-8. Locating the Intersection Point of Three Circles

sured. The correction weight required is determined by Equation (6.2).

$$\text{Correction Weight} = \text{Trial Weight} \times O/T \qquad (6.2)$$

The angle formed by the T vector is the location of the correction weight. This process can be repeated as often as required until acceptable results are obtained. Generally, two balancing runs should provide good results. As with all maintenance practices, careful attention to small details will result in big dividends.

Chapter 7

Misalignment of Machine Shafts

INTRODUCTION

Vibration due to misalignment is almost as common as vibration due to unbalance. The misalignment can be internal, that is misaligned bearings, or external such as two pieces of equipment that are coupled together. It takes a precise method of alignment to assure these parts are aligned so that no forces are generated due to their misalignment.

In the case of an anti-friction bearing being misaligned with its shaft, the vibration will exist in both the radial and axial directions. Figure 7-1 illustrates an anti-friction bearing misaligned to the shaft. Proper installation of the bearing will eliminate this vibration.

Sleeve bearings must have an accompanying unbalance to exhibit classical misalignment vibration characteristics. Again, if an unbalance exists, the shaft will have both radial and axial vibrations. These vibration characteristics are due to the bearings'

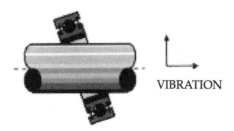

VIBRATION

Figure 7-1. Anti-Friction Bearing with Misalignment

reaction to the unbalance. The real source of these characteristics is thus the unbalance. The amplitude of both the radial and axial vibrations will be reduced when the rotor is balanced.

Figure 7-2 depicts misalignment of a sleeve-type bearing to its shaft.

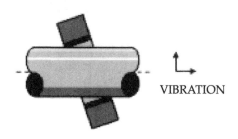

VIBRATION

Figure 7-2. Sleeve Bearing with Misalignment

TYPES OF MISALIGNMENT

Misalignment in couplings can manifest itself in three basic ways. The first is parallel offset as shown in Figure 7-3. In this type of misalignment, the two shafts can be offset vertically, horizontally, or a combination of both.

This type of misalignment is often found in small equipment that has a common base plate, and the equipment is doweled to the base plate. Many manufacturers dowel the equipment in this fashion, but it is always wise to check for proper alignment. This often requires discarding the dowels.

The second type is angular misalignment, and is depicted in Figure 7-4. In this type of alignment, the angularity again can be in the vertical plane, the horizontal plane, or both. Misalignment will almost always be in both the vertical and the horizontal planes.

In most cases, misalignment in couplings will not occur as either a pure parallel offset or as a pure angular misalignment, but rather as a combination of both. This is the third type of misalignment and is shown in Figure 7-5.

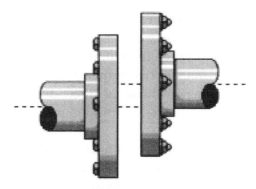

Figure 7-3. Coupling with Parallel Offset Misalignment

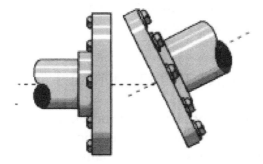

Figure 7-4. Coupling with Angular Misalignment

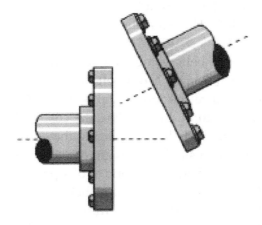

Figure 7-5. Coupling with Most Common Form of Misalignment.

A bent shaft exhibits vibration characteristics very much like those of misalignment. However, it is easily detected when alignment attempts fail to reduce the vibration amplitude.

Misalignment results in two primary forces, one in the radial direction, and the other in the axial direction. This is true even with flexible couplings where the misalignment is within the tolerances of the coupling.

As with unbalance, the amplitude of the vibration will increase as the problem is worsens. Therefore, the more misalignment, the more vibration. The key to detection of misalignment or a bent shaft is the high axial vibration.

In general, the vibration is at 1 × rpm, but when the misalignment is severe, second order (2 × rpm) and even third order (3 × rpm) vibrations can occur.

MISALIGNMENT IN BELT DRIVES

Misalignment of sheaves sprockets and V-belt drives also produces these vibration characteristics. This type of misalignment results in severe wear to chains, V-belts, sprockets, and sheaves. Misalignment in V-belt drives that employ multiple belts can create a host of other problems as well. Misalignment will alter the tension of the different belts, so some may slip and others may be overloaded. This can obviously lead to premature belt failures.

Generally, these belts come as a set and thus their replacement cost is normally high. Multi-belt systems should always be replaced as a set, since the individual belts are matched to each other at the manufacturer. In addition, the cost of down time and labor to change a single belt vs. replacing the entire set is usually nil. However, if only a single belt were replaced and another belt in the set were to fail in a few hours of operation, these costs would far offset the price of the set of belts.

One method of detection, is to draw a chalk mark across the set of belts when the unit is off. Using a strobe light, the frequency is adjusted to the rpm of the belts, and the different speeds of the belts can be seen.

As an alternate method, operate the unit for a very short time (bump the motor) and inspect the location of the marks on the belts. If they have departed from each other, check for possible misalignment. Figures 7-6 and 7-7, show exaggerated misalignment in V-belt drives.

The amount of misalignment shown in Figure 7-6 would result in the drive belts being thrown off the pulleys. However, it can be seen from the figures that misalignment can result in vibration characteristics similar to misalignment in shafts.

Too often, it is thought that belt drives are self-aligning and thus little to no attention is paid to their alignment. This is simply

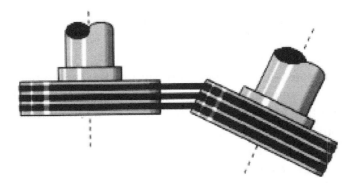

Figure 7-6. Angular Misalignment in Belt Drives.

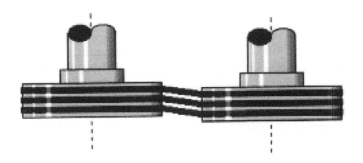

Figure 7-7. Offset Misalignment in Belt Drives

not the case. This type of misalignment will manifest itself into the same failures of bearings as shaft misalignment.

Figure 7-8 shows why using a straightedge to align pulleys or sheaves can lead to errors. Note that the two pulleys have a different face thickness. Aligning these pulleys with a straightedge causes the sheave centers to be offset. Be sure to measure the thickness of any pulleys that will be aligned using this method. It is far better to use the pulleys' center-lines as a reference.

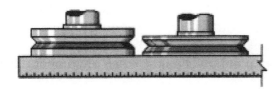

Figure 7-8. Errors in Using Straightedge to Align Pulleys

Regardless how V-belt and chain drives are aligned, the amount of axial vibration is a sure sign of the preciseness of the alignment. In general, when the axial vibration exceeds 1/2 the highest vertical or horizontal vibration, misalignment or a bent shaft should be suspected.

To determine if the cause of vibration is related to the belt itself, the rpm of the belt needs to be determined. The following formula determines the belt rpm.

rpm = 3.14159 × Pulley Diameter/Belt Length   (7.1)

If the vibrations are at belt rpm or a multiple of belt rpm, the belt(s) should be suspect. Belt defects can be in the form of missing material, uneven widths, lumps on the belt, cracks, hard spots, soft spots, and others. Regardless, these will show up at belt rpm. Multi-belt applications may exhibit frequencies that do not coincide with belt rpm, since several belts could have defects at different locations.

Belt tension is very important in assuring the long life of the belts and equipment. Although belts that are too loose may not be able to transmit all the required power, there are other problems as well. Belt tension directly affects the natural frequency of the belts. Belts that are either too tight or too loose can go into resonance. This in turn will cause a vibration which can be harmful to the bearings and seals.

PHASE ANGLE RELATIONS

Although phase analysis is useful in determining whether a bent shaft or misalignment is the source of the vibration, it is not foolproof. The easiest solution is to check the alignment with dial indicators. Precise methods are discussed in following chapters. It is best to check the alignment with the equipment at operating temperature to assure the problem is not due to thermal growth.

The first step in using phase angles as a diagnostic tool is to locate a reference point on the shaft or coupling that is easily visible throughout the entire 360 degrees of rotation. Often, the keyway in the shaft serves this purpose well.

A vibration analyzer with a strobe light attachment is required to take these readings. The direction of the vibration pickup must not be changed from reading to reading. Changes in the direction in which readings are taken will cause a phase shift.

The types of readings taken (displacement, velocity, acceleration) cannot be changed when taking phase readings. To do so will change the phase angle.

Since the objective is to determine a bent shaft from misalignment, the strobe light should be set to trigger on the vibration, but the filter should be used to assure the vibration source is a 1 × rpm. The filter should not be re-tuned during the other phase readings.

For the initial test, four readings around the bearing should be taken. Typically, the 12, 3, 6, and 9 o'clock positions should be used. The end of the shaft can be viewed as a clock face so that the phase mark can be referenced to the points of the clock. A reading at 3:00 o'clock would be 90 degrees from the top; a reading of 7:00

would be 210 degrees from the top, etc.

Figure 7-9 illustrates the imaginary clock face around a bearing.

It is not important as to where the original phase mark appears, but rather the change in position as readings are taken at various locations. When this test is being performed for analysis of potential misalignment, the basic clock hours are close enough. However, when balancing, a polar graph paper taped to the machine may be useful in increasing the accuracy of your measurements.

In Figure 7-10, the four points to measure phase angle on a bearing are located. Starting at point A, locate the keyway or other mark on the shaft and note its phase angle. Repeat for the remaining three points. If the phase mark remains relatively unchanged, the bearing is moving in phase—that is, the entire bearing is moving in and out at the same time.

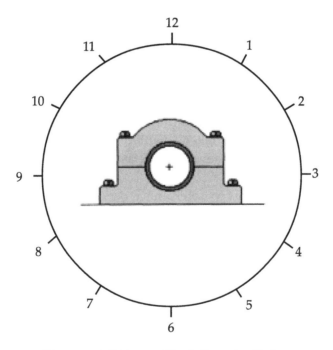

Figure 7-9. Using a Clock Face for Reference

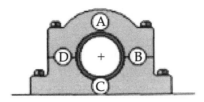

Figure 7-10. Locating Four Reference Points

The next step is to measure the bearing across the coupling. Often, this will require the pickup to be placed in the opposite direction. If the pickup is reversed, the new phase readings must be adjusted by 180 degrees.

In Figure 7-11, the direction of the vibration pickup for each bearing location is shown. Assuming the initial readings were taken at point B (to use the keyway as a reference point), readings taken at points A and C would need to be corrected by 180 degrees.

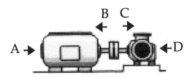

Figure 7-11. Direction of Vibration Pickup at Each Bearing

In general, if the readings taken at points A and B are close to being the same phase angle and the readings of points C and D are close to each other; and the points B and C are 180 degrees out of phase, the problem is usually between the two bearings. This could be a coupling problem or a strong, but not conclusive, indication of misalignment.

It is always best to check the alignment and inspect the coupling at this point. Refer to the chapters on alignment for inspections to be made on the coupling.

NOTE ON ELECTRIC MOTORS

In general, the electric motor is considered as the movable machine when performing alignment of equipment trains. Often this is due to constraints caused by piping or other connections to the driven equipment. Usually this creates little or no problem, since the electrical connections to the motor are most often terminated with a short piece of flex conduit.

However, some potential problems can occur when aligning the motor to a stationary piece of equipment. In some cases, the motor may be aligned in such a manner that its magnetic center and its gravitational center do not coincide.

This internal misalignment of centers can cause the motor's rotor to pulse as it operates. This is due to its hunting for the magnetic center while being pulled to its normal center of gravity.

Motor Rotor Operating
In Its Magnetic Center

Figure 7-12. Rotor Operating in its Magnetic Center

The motor rotor can be pulled from its magnetic center by the equipment it is coupled to. That is, the driven equipment may produce a force across the coupling that attempts to either push or pull the motor rotor from its magnetic center. In either case, the magnetic field will try to force the rotor back to its magnetic center, while the other force tries to force it away from the magnetic center. This force imbalance often results in the motor rotor surging back and forth, pounding the thrust bearing of the driven equipment.

This problem is easily identified, and easily cured. The motor should be stopped and disconnected from the coupling of the driven equipment. The motor is started and allowed to reach full

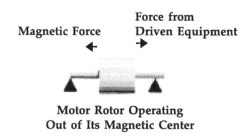

Motor Rotor Operating
Out of Its Magnetic Center

☐Figure 7-13. Rotor Operating dfits Magnetic Center

speed. Using a piece of chalk or a felt-tip marker, mark the shaft where it protrudes from the bearing housing. Shut off the motor and observe it as it coasts to a stop. Note the movement (if any) of the mark on the shaft. If the mark moves either in or out, the magnetic center and the gravitational center are not the same.

The bearings or the assembly of the motor should be inspected. Unless specifically designed for operation other than level, the motor should be placed in service in a nearly level position. If it is far off of level, the motor should be leveled and corrections made to the position of the driven equipment so that proper alignment can be achieved without moving the motor from a near-level position.

Care should be exercised to assure that the re-assembly of the coupling between the motor and the driven equipment leaves the motor shaft in its magnetic center position. Couplings with float should be assembled so that the motor shaft has its mark in the center of the movement.

After the equipment is properly assembled and aligned, re-start the equipment an observe the position of the motor's marked shaft.

SUMMARY

In general, the best indicator of misalignment is the high axial vibration. As a guide, anytime the axial vibration is at least 70% of the radial vibration (either horizontal or vertical), misalignment should be suspected.

Phase analysis is useful in the final determination of the exact cause of the high axial vibration. If phase readings taken on both outboard bearings of a machine indicate that the bearing housings are vibrating in phase with one another; and the same is true for the second machine; and a notable phase shift is seen when taking readings across the coupling, the source of the axial vibration is somewhere in between. The coupling or misalignment is obviously suspect. If a phase shift is noted between the two bearings of the same machine, a bent shaft should be suspect.

In following chapters, several precise methods for alignment of machine shafts are presented. In addition, several supplementary considerations for alignment are discussed. These include alignment of equipment trains, alignment of equipment with drive shafts and alignment of U-joint types of couplings. Thermal growth considerations are also presented.

Chapter 8

Advanced Machine Alignment

INTRODUCTION

Precision alignment of equipment shafts to one another is critical in extending the useful life of various machine components, and in preventing catastrophic failures. The two most common sources of machine vibration are misalignment and unbalance. Often these two sources will account for over 90% of all machine problems occurring in any given plant.

During the alignment process, many sources of machine unbalance can be detected and corrected. Thus, a complete understanding of precision alignment and an understanding of the sources of machine unbalance are critical skills to the plant repairman. The topic of unbalance has been covered, but will be discussed during the alignment presentations.

The first method of alignment that will be presented is an advanced version of the commonly known Rim and Face method. Both graphical and analytical solutions to the alignment process will be presented. Thermal growth and other machine considerations will be discussed in detail. In the following chapter, the reverse indicator method of precision alignment will be discussed in detail.

Often equipment manufacturers provide tolerances for alignment of their equipment. Usually this is in the form of +/- so many mils parallel offset and +/- so many mils angularity. The very word tolerance indicates a willingness to accept or tolerate some misalignment. In this text, methods will be detailed to attempt to align equipment to exactly zero tolerance.

In addition, methods for compensating for thermal growth, and for checking and correcting running alignment will be discussed. This will assure the equipment is precisely aligned once it has stabilized at operating temperatures.

OVERVIEW

Misalignment of two machine shafts can be parallel offset, angular, or a combination of both. In addition, the misalignment can be in the horizontal plane, the vertical plane, or both planes. Generally, most situations will be a combination of all the above. However, when the problem is attacked one step at a time, the solution is relatively simple.

Regardless of the method of alignment to be employed, it is important to examine the equipment to determine which machine will remain stationary, and which machine(s) will be moved. This selection is often influenced by such things as connected piping or fan blades, which must remain, centered in a shroud. In general, the driver is usually the adjustable piece of equipment, since it generally has less constraints attached, such as piping. Once the stationary piece of equipment has been identified, several important checks must be performed to assure a proper alignment.

First, pumps and compressors, especially those with a maintenance history of bearing or seal problems, should have their piping taken loose to assure it is not in a strain. If it requires excessive force to re-connect, piping modifications should be considered. Here is another example for the need to keep and maintain informative maintenance records. Once the piping has been inspected, there will be no need to recheck its alignment unless some system changes are made. Later, it will become obvious that other information about the alignment will be beneficial in future alignment attempts.

Fans should be inspected to assure they are centered within their shrouds. Aerodynamic forces resulting from improper centering can lead to excessive vibrations and premature component failure. In addition, all blades should be inspected for buildup of foreign material, and to assure they are all set to the same and correct pitch.

Next, the stationary machine should be checked to assure it is near level. This is more critical in certain types of equipment, but should be considered for all rotating machinery. This will assure that the adjustable piece of equipment starts out on a level plane, and should allow for a minimum of shims being required to bring the two pieces of equipment into alignment.

ELECTRIC MOTORS

In the case of electric motors, being out of level can result in the rotor surging along its axis. Since some electric motors do not have thrust bearings, the resulting pounding must be absorbed in the coupling or the thrust bearing of the driven equipment. This surging results from the gravitational center of the rotor being different than the magnetic center. The rotor continually hunts for the position it wants to run in. This manifests itself in high axial vibrations. The phase and frequency of these vibrations may or may not prove to be synchronous.

This condition is not difficult to determine. Use chalk or a felt marker and draw a line where the motor's shaft exits the bearing housing. Disconnect and secure the coupling, and briefly run the motor. Allow the motor to coast to a stop undisturbed. Inspect the mark. It should be in the same location as when the motor was running. If it has moved either in or out, there may be a problem with the gravitational center. In the case where the motor was moved to align it with the stationary equipment, the stationary equipment must be re-leveled.

SHIMS

To assure a precision alignment, the feet or hold-down bolt pads of the equipment must be clean and free from rust, dirt, and oil. In addition, it is wise to remove all old shims and inspect them. Often, equipment has been aligned using any material available, such as soft drink cans, galvanized sheet metal, tin cans, and other scrap metal. Obviously, this is undesirable and must be discarded, and the proper shim material used. Regardless of the type of shim material to be used, all shims must be miked to verify their thickness. All cut or stamped out shims should be carefully inspected for burrs that can distort their thickness. The use of brass or some of the newer plastic shims is highly recommended.

One final caution about shims: The use of excessive numbers of shims can cause uneven crush each time the machine hold-down bolts are tightened. This will result in different indicator readings and create many problems in the alignment process. It is therefore recommended that no more than three shims are placed under any machine foot. This may require machining a plate of a given thickness for each foot, but the dividends will be proper alignment and greatly extended equipment life.

Once the shims have been cleaned (or replaced) and the machine feet and base plate have been cleaned, re-torque the hold down bolts on the stationary machine to the proper tension. The use of a torque wrench is highly recommended to assure not only the proper preload, but to assure all bolts are tightened the same, and that any subsequent re-tightening will place the same amount of crush on the shims.

SOFT FEET

The first step is to assure the stationary machine is sitting squarely on its foundation or base plate, and is supported equally on all feet. A condition known as soft feet can cause internal misalignment of the machine and lead to premature failures. Checking for soft feet is a relatively simple task. Once the shims have been removed, the machine feet have been cleaned and inspected, and all hold-down bolts are properly torqued, loosen one hold down bolt at a time. Use either a feeler gauge or a properly mounted dial indicator to check the amount of movement or the gap under the foot.

Re-torque the hold-down bolt and move to the next one. Draw a simple sketch of the machine, and record the movement or gap for each foot. All feet should exhibit about the same movement or gap. If the difference is more than 2 mils, add the proper amount of shims under the feet with the excessive movement, so that all are the same. Once this has been completed, the work on the stationary machine is completed, except for some checks for potential sources of unbalance. This procedure should now be performed on the adjustable machine. See the appendix section for additional details.

Regardless of the alignment method to be employed, it is highly recommended that the vertical alignment be corrected first. This will allow the machine to be moved horizontally without disturbing the vertical alignment. If the horizontal alignment were corrected first, moving the machine in the vertical plane would most likely disrupt the horizontal alignment.

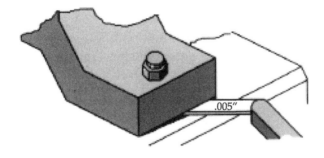

Figure 8-1. Using a Feeler Gauge to Measure for Soft Feet

TORQUING

It is imperative that hold-down bolts be torqued each time they are tightened. This not only assures the bolts are not too loose or too tight, but also will provide a constant crush on the shims. Uneven tightening or tightening to different clamping forces can cause false readings during an alignment process.

Each grade of fastener has specific limits based on the strength of the material it is manufactured from. In general, industrial fasteners used as hold-down bolts, will normally be either grade 5 or grade 8. These fasteners are identified by the number of radial lines present on their heads. Grades 2 and below have no markings. Grade 5 has three lines and grade 8 has six lines.

GRADE 2   GRADE 5   GRADE 8

Figure 8-2. Determining Bolt Grade

Due to their poor characteristics, grade 2 bolts are seldom used in industrial applications. It is very important that the fastener is properly identified and that it is properly torqued to the specifications for that particular bolt.

As an example, a 1-1/4 inch bolt with 7 threads per inch has a tensile stress area of .9646 square inches. Table 8-1 reflects the various properties for grade 2, 5 and 8 bolts.

Table 8-1. Properties of Various Grade Fasteners

GRADE	CLAMP FORCE (lbs.)	TORQUE (Ft.-lbs.)
2	19,176 – 28,719	300 - 450
5	43,023 – 64,535	672 – 1,008
8	69,768 – 104,652	1,090 – 1,635

Table 8-1 reflects the danger of selecting the wrong grade fastener for an application. If a grade 2 fastener were used, and torqued as if it were a grade 5 fastener, it would fail prior to reaching the minimum torque value of the grade 5. If the grade 2 fastener were torqued to its maximum, it would produce a clamping force of only 67% of the minimum clamping force of the grade 5 fastener.

Most often, fastener manufacturers state the recommended torque value for clean, dry threads. Any lubricant or debris on the threads will greatly effect the torque value.

A very small amount of oil on the threads will reduce the friction significantly—enough to reduce the recommended torque by about 15% to 25%. Some of the newer Teflon- or Molybdenum-based dry type lubricants can reduce the torque required by as much as 50%. Often, zinc-plated fasteners can be found in industrial applications. This plating will reduce the amount of required torque by 15%.

Table 8-2 illustrates the tensile strength for various grades of fasteners.

Table 8-2. Properties of Various Grade Fasteners

GRADE	MATERIAL	TENSILE STRENGTH (psi)
2	Low carbon	74,000
5	Medium - carbon, tempered	120,000
8	Medium - carbon, tempered & quenched	150,000

If a fastener is subjected to excessive torque, which stresses the fastener beyond its elastic limit, there is a necking down of the material. That is, the bolt has a permanent offset or stretch and has a reduced cross-sectional area.

Figure 8-3. Reduced Cross-sectional Area Due to Over Torquing

Although bolts often fail in the thread area, since this is the minimum cross-sectional area and also an area of stress concentration, Figure 8-3 shows an exaggerated effect of over torquing.

Using 72% of the minimum tensile strength as a limit, the 1-1/4 inch grade 8 fastener with a cross-sectional area of .9646 and a tensile strength of 150,000 psi would safely produce a clamping force of 104,209 pounds. If this fastener were stressed beyond its elastic limit such that there was a reduction in its diameter of .15 inches, its effective cross-sectional area would be reduced to .7212 square inches, and its clamping force limit would be reduced to 77,844 pounds, or 74.7% of its original effectiveness.

A table of recommended torque values is provided in the appendix section. Always use the proper grade fastener and follow the equipment manufacturer's recommended torque values. The table of torque values in the appendix is provided as a general guide only.

COUPLINGS

When utilizing this method of alignment, it is not necessary to disassemble or remove the coupling. However, a complete inspection of the coupling is highly recommended. The coupling and its associated components are potential sources for machine unbalance. Refer to Figure 2-8 for potential sources of coupling unbalance.

First, inspect all the coupling bolts to assure they are all the same length, diameter and grade, and that they all have the same number of nuts and washers. Second, inspect the keys and keyways of the coupling, driver, and driven equipment. Most equipment manufacturers balance their rotating assemblies with a full half key in place. That is they completely fill the key way with key material, but only to the surface of the shaft or bore. If equipment is assembled with a full-length, full-height key or a full-height partial-length key, unbalance may be introduced into the machine.

Although this may seem like splitting hairs, recall from the examples in Chapter 2 that it is the seemingly minor things that can cause major problems. Large and obvious defects are usually corrected when they are detected. However, the smaller seemingly insignificant things are too often ignored.

THE ALIGNMENT FIXTURE AND BAR SAG

One of the first steps in any alignment process is to determine how to set up a fixture that will allow you to take the required measurements. In aligning two shafts, the readings are taken from the stationary shaft to the adjustable shaft, across the coupling. *Always mount the fixture to read from the STATIONARY equipment to the ADJUSTABLE equipment!*

Once you have set up a suitable fixture to allow attaching the two dial indicators to read a rim and a face reading across the coupling, the fixture should be removed without disturbing the distances being measured. Prior to removing the fixtures, their exact location should be marked on the shafts, so they can be reinstalled in the same location.

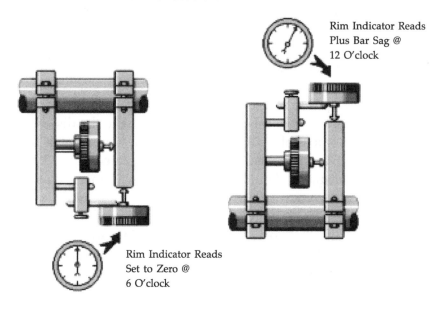

Rim Indicator Reads
Plus Bar Sag @
12 O'clock

Rim Indicator Reads
Set to Zero @
6 O'clock

Figure 8-4. Measuring Bar Sag

The fixture is now attached to a rigid pipe or shaft, and the bar sag can be measured. Bar sag is very important in any precision alignment technique. Not compensating for bar sag can create errors sufficient to void the best alignment process. It should be noted that the error created by not properly compensating for bar sag is multiplied over the length of the machine. Thus an error of 2.5 mils over a 12" distance becomes an error of 11.7 mils on a machine foot 56" away.

To measure the bar sag, the pipe with the fixture attached is rotated so that the rim indicator is in the bottom or 6 o'clock position. The indicator is then set to indicate zero.

The entire fixture and pipe are now rotated to the top or 12 o'clock position and the indicator read. It will indicate a plus (+) reading which is the bar sag. The bar sag must be recorded on the alignment form for future reference.

The fixture should now be re-attached to the machine in the 12 o'clock position. By leaving the rim indicator set to the (+) bar sag reading, no further compensation will be required for bar sag. Always assure the rim indicator is set to the (+) bar sag when it is in the 12 o'clock position.

BASIC MACHINE MEASUREMENTS

The next step is to measure the machine distances. As with all measurements, care should be exercised to measure the distances as accurately as possible. The most critical measurement is the diameter traced by the face indicator as it is rotated around the machine. This distance should be measured to within ±1/8 of an inch.

Normally, it is easier to rotate the fixture so that the face indicator is in the 3 o'clock or 6 o'clock position, then measure from the center of the shaft to the center of the indicator stem, and multiply the measured distance by two. This distance becomes the A distance on the alignment work sheets.

Next the distances to the hold-down bolts are measured.

Generally, these distances should be measured to the nearest 1/4 of an inch. These distances are measured from the center of the stem of the rim indicator to the center of the hold-down bolt.

A carpenter's square is useful in transferring these measurements to the machine centerline, where the measurements are more easily taken. The distance to the adjustable machine's near foot is the B distance and the distance to its far foot is the C distance.

Some machines have more than four feet. In this event, measure and record the distance from the near foot to the remaining feet that are located between the near and far feet, and record the values and assign a label to each additional foot location.

The distances to the feet of the stationary machine should also be measured for future reference, or in the event that thermal growth will be considered. The topic of thermal growth will be covered later.

All of the measured distances should be recorded on a simple sketch of the machine, or on forms similar to those contained in the appendix. These distances will be scaled and used for the graphical solution to the alignment. They will also be used in the simplified calculator method and can be input into a computer alignment program.

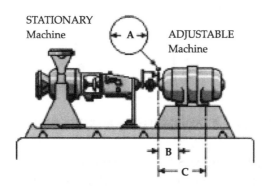

NOTE: "A" is the diameter traced by the face indicator.

Figure 8-5. Basic Machine Measurements

ABOUT INDICATOR READINGS

Since two points determine a straight line, the objective in any alignment process is to locate two points on the centerline of the adjustable machine, to determine its relative position. Once these two points are located, with respect to the stationary machine centerline, the required movement to align the two shafts is easily determined.

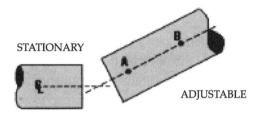

Figure 8-6. Locating the Centerline of the Adjustable Shaft

Locating the two points can be accomplished by either obtaining two rim readings, or by obtaining a rim reading and a face reading. The two rim reading method is referred to as the reverse indicator method, and the latter is referred to as the rim and face method.

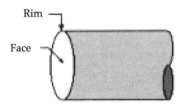

Figure 8-7. Nomenclature of Shaft Readings

Referring to Figure 8-8, the distance b is the height of the bracket above the centerline of the stationary shaft. At point 0, the dial indicator would be set to the plus bar sag, but ignoring bar sag, it is set to zero.

As the shafts are rotated 180 degrees, the vertical distance traveled by the indicator stem is 2b. The indicator is now at point 1. The distance 1 to 2 is the total reading obtained from the dial indicator.

Since the total indicator reading (TIR) was a result of a vertical change of 2b, a vertical change of b must equal a reading of the TIR/2. This is why all rim and bore readings must be divided by two.

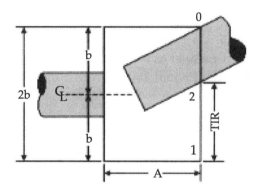

Figure 8-8. Geometry of a Rim Reading

The nature of a dial indicator is such that as the stem moves out from the body, the dial reads in the negative direction. In Figure 8-8, the indicator stem moves outward from point 1 to point 2 recording a negative reading. Since the indicator is actually mounted on the stationary shaft, the reading is stating that the stationary shaft is lower than the adjustable shaft. Thus, if the stationary shaft is lower than the adjustable shaft, the adjustable shaft must be higher than the stationary shaft.

Anytime a reading is changed from one shaft to the other, its algebraic sign must be changed. Since the reading is taken on the adjustable shaft, and the position of the adjustable shaft with respect to the stationary shaft is required, its sign is changed. This is why rim readings have their algebraic sign changed.

These methods of correcting the indicator readings will be used in both the rim and face and reverse indicator alignment processes. They are discussed in more detail later in this and the following chapters.

In Figure 8-9, the distance A is the diameter traced by the face indicator, reading from the stationary shaft to the adjustable shaft. The indicator is set to zero at point 1 and rotated 180 degrees to obtain a reading at point 2. The resulting indicator reading is the distance 2 to 3. The triangle 1-2-3 can now be constructed. Note that the angles c are 90 degrees, and since the sum of the angles of a triangle equals 180 degrees, angles a plus b must equal 90 degrees.

If a line 4 to 5 were laid out with a length A' which is equal in length to A, and parallel to the line 2 to 3, the sum of the angles formed at point 4 must equal 180 degrees. Since the angle formed from lines 1 to 2 and 4 to 5 is 90 degrees, angle c, and the angle formed from lines 1 to 2 and 4 to 6 must be equal to angle a, and the angle formed by lines 4 to 5 and 4 to 6 must equal angle b. Therefore, the two triangles 1-2-3 and 4-5-6 must be equal.

This converts a face reading into a rim reading a distance A away from the indicator stem along the adjustable machine centerline. Note that since this reading was taken over the distance A, and converted to a rim reading A distance along the centerline, this reading is not divided by two.

Like all readings, when referenced to the stationary shaft, its algebraic sign must be changed.

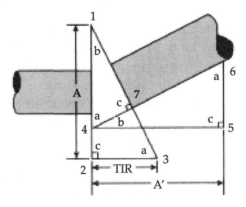

Figure 8-9. The Geometry of the Face Reading

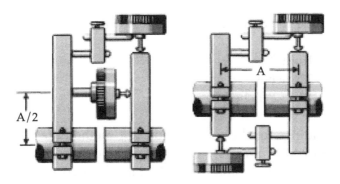

Figure 8-10. Basic Alignment Fixtures

A major difference in these two methods of alignment is that the reverse indicator method takes one reading from the stationary shaft to the adjustable shaft, and one reading from the adjustable shaft that is referenced back to the stationary shaft. The rim and face method takes two readings from the stationary shaft to the adjustable shaft.

This principal difference determines the way the readings are handled, and the subsequent plotting and calculating of the required movement. The reverse indicator method requires that both readings are divided by two, since both are rim readings, and that the adjustable indicator have its algebraic sign changed to reference it back to the stationary shaft. The rim and face method requires that both readings have their algebraic signs changed to reference back to the stationary shaft, and that the rim reading be divided by two.

The rim and face method requires that the face reading be referenced from the rim reading, and is plotted a distance A away from the rim point, and is marked off vertically from that point and not from the centerline of the stationary shaft.

Example 8-1

The alignment is to be measured on a machine by both the reverse indicator and the rim and face methods. The fixtures are so located and assembled that the A, B, and C distances will remain the same for both sets of measurements.

A = 10 inches.
B = 22 inches.
C = 45-1/2 inches.

The readings for the reverse indicator method are S = minus four (–4) and M = zero (0). The rim and face readings are R = minus 4 (–4) and F = plus two (+2). Plot the two methods and assure the same movements are indicated.

For the reverse indicator method, both readings are divided by two and the adjustable machine indicator has its algebraic sign changed. This yields S = 2 and M = 0. These two points are marked off on a graph A distance apart. The two points are connected, and this line is extended to intersect two lines representing the inboard and outboard foot locations. The result is to add 4.5 mils to the inboard foot and 13 mils to the outboard foot.

For the rim and face method, the rim reading is treated exactly as the adjustable reading in the reverse indicator method. That is, it is divided by two and has its algebraic sign changed. The face reading simply has its algebraic sign changed. This results in R = plus two (+2) and F = minus two (–2). The corrected rim reading is plotted the same as the stationary reading in the reverse indicator method. A horizontal line is now constructed to

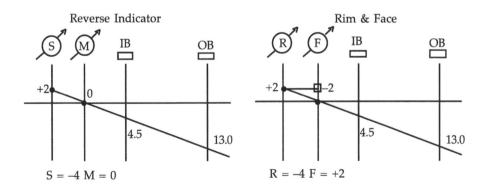

Figure 8-11. Plotting the Alignment Results

intersect the face line a distance A away. From this point, the corrected face reading is measured off. Again a line is constructed passing through these two points and extending through the two foot locations. The result is identical to the result found in the reverse indicator method: add 4.5 mils to the inboard foot and 13 mils to the outboard foot.

Both of these methods will be explored in detail in this and the subsequent chapter.

TAKING INDICATOR READINGS

Once soft feet are corrected, machine measurements taken, the fixture assembled, and bar sag determined, an initial set of readings can be taken. Be sure to always rotate both shafts together (in the event the coupling is not made up), and always rotate the machine in the direction of normal operation. This is especially true of equipment that has gears or tilt pad thrust bearings.

Never back up to get a reading. If you rotate the fixture too far, continue around in the same direction until it lines up. For the vertical alignment readings, the rim indicator is set to the (+) sag and the face indicator to zero (0) in the 12 o'clock position. The shafts are rotated to the 6 o'clock position and a set of readings is taken. To assure the information is correct, continue to rotate the shafts until they return to the 12 o'clock position, where the indicators should return to their original readings. If the readings do not repeat within 1/2 mil, the fixture is not securely mounted or is not sufficiently ridged.

Prior to any actual alignment corrections, several things must be performed. As with the stationary machine, the adjustable machine must have its shims, feet, and base plate inspected and cleaned. Checks for soft feet should be made as was done for the stationary machine. On the adjustable machine, this should be performed with all old shims removed. The amount of shims under each foot should be measured, verified with a micrometer, and recorded. Generally, these old shims should be discarded, and the alignment process started from scratch. Once a solid alignment program is in place in your plant, this may not be required.

VERTICAL ALIGNMENT

When dealing with cut shims (cut from stock), or stamped shims, extreme care should be exercised to assure no burrs exist. All shims and shim stock to be used in the alignment process must be measured with a micrometer to verify their thickness. In addition, no more than three shims should be placed under any machine foot. This may require obtaining special shim stock, or perhaps machining spacers. However, this will eliminate most of the possibility of uneven shim crush that can destroy any alignment attempt.

Properly torque down all hold-down bolts and take an initial set of readings as outlined previously. As a word of caution, observe the dial indicators to assure you know which way they moved. Often, they will start in one direction and then reverse and read the opposite. A large positive reading can be mistaken for a small negative reading. Once again, return the indicators to the starting position to assure the readings repeat. If the numbers repeat and you rotated the equipment in its normal operational direction, and you did not go past the 6 o'clock mark and reverse, you are ready to determine the required vertical corrections.

Some people find the graphical method easier to use than the calculator, while others prefer the calculator to the graph. However, it is highly recommended that both methods be used to assure the proper corrections are made. Both methods require only about 2 to 3 minutes each, once you have mastered them, or at least lost the fear of them.

Both methods are capable of accuracy of better than $1/10^{th}$ mil, but no one would attempt such an alignment; besides, $1/10^{th}$ shims are hard to find. The point is that accuracy to within ±1 mil will assure an exact alignment and long equipment life. Be sure that answers obtained from the graphical method and calculator method are the same (when rounded off to the nearest mil). Generally, the calculator method will provide the exact answer, while the human eye tends to round off the graph, or the pencil point or straightedge slips slightly.

Regardless of the method employed, when the proper shims are added and the machine properly torqued, the indicators should zero in the 6 o'clock position. Often, due to crush, torque, dirt and other discrepancies, more than one alignment attempt will be required to obtain zeros.

THE GRAPHICAL METHOD

The first step in the graphical solution is selecting and laying out the vertical and horizontal scales on the graph. Note that the graph is a Cartesian coordinate system as studied earlier. Generally, the vertical scale can be set at one square equals one mil. Unless the machine is severely misaligned this should be adequate. The horizontal scale should be determined from the machine measurements. If thermal growth is to be considered, the scale must be such that both the stationary and adjustable machines will fit on the graph.

For instance, if the distance from the rim to the stationary machine's outboard foot were 48 inches, and the distance from the rim indicator stem to the outboard foot of the adjustable machine were 56, then a scale should be determined so that 104 inches would fit on the graph. Assuming that the graph paper is 8-1/2 by 11 and that the horizontal scale will be laid out on the 11-inch side, a scale of one inch equals 10 inches could be used.

Starting at the left edge of the graph, a vertical line is drawn near the edge to represent the outboard foot of the stationary machine. A symbol representing a machine foot is drawn at the top of the line, and labeled O.B. for outboard foot. A horizontal line is drawn halfway up the graph, which represents the final position of both machine shafts when they are aligned to each other. This line is labeled the final alignment line. Note that if thermal growth is not to be considered, this line represents the present location of the stationary machine's shaft centerline.

Next, using the scale factor (in this case 1:10), measure from the first vertical line to the location of the rim indicator and draw another line. This line is labeled the rim indicator line, and a symbol is also drawn at its top. The distance from the outboard foot to the stem of the rim indicator is 56 inches ("E" distance), and would be drawn 5.6 inches from the first line.

The location of the other machine feet can now be determined from the position of the rim indicator line. Measure off and label the other three machine feet using the B, C and D distances, and the scale factor. Finally, the distance A is measured off from the rim indicator line to the right, and its vertical line is drawn and labeled face indicator. Once again, symbols are used to assist in identifying each of these lines as shown in Figure 8-12. The alignment data can now be plotted on the graph.

First, the indicator readings must have their algebraic signs changed. This is necessary to convert the readings from the adjustable machine to the stationary machine. (If X is bigger than Y, then Y must be less than X.) The rim reading is divided by two, because

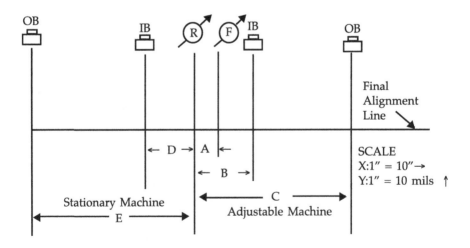

Figure 8-12. Rim & Face Graphical Layout

the nature of a rim reading is such that it measures the offset at the top and the bottom (double the actual misalignment). This value can now be plotted on the line representing the rim indicator.

Continuing, if a reading of minus eight (–8) mils is recorded on the rim indicator, first the algebraic sign is changed and the value is divided by two. This yields plus four (+) 4 mils. Since the vertical scale was set at one square equals one mil, the point is located four squares above the horizontal final alignment line, and on the rim indicator line. This point is marked and circled.

A horizontal line is now drawn from this point to the same point on the line representing the face indicator. The point on the face indicator line has a box drawn around it. The recorded value of the face indicator is now counted off from this point.

Remember, positive numbers are measured up and negative numbers are measured down. Assuming that in this case the face indicator reading was a plus two (+2), changing the sign yields a minus two (–2). A point two squares below the horizontal line just drawn represents the face indicator reading. This point is marked and a circle is drawn around it.

Using a long straightedge, align it so a line can be drawn through the two-circled points. This line should extend to the right far enough to cross the adjustable outboard foot line. Label this line the center of the adjustable machine shaft.

Assume in this case that the A distance is 10 inches and the B distance is 18 inches. Where this line crosses the feet of the adjustable machine indicates the amount of shims to be added or removed. If the adjustable machine centerline crosses the foot line above the final alignment line, shims must be removed; if it is below, shims must be added.

The amount of shims to be removed or added is determined by counting the squares (up or down) from the final alignment line to the adjustable machine centerline at each of the foot locations. In our example, the inboard foot would require .5 mil to be removed and the outboard would require 5.5 mils to be added. Since we are only to deal with whole mils, the inboard foot would be left as is, and the outboard foot would have 5 mils of shims added.

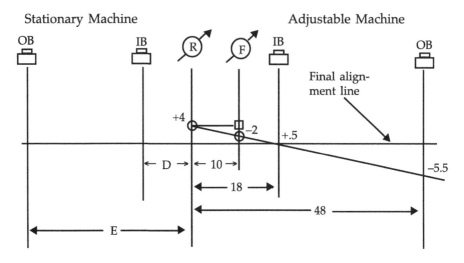

Figure 8-13. Rim & Face Graphical Solution

CALCULATOR METHOD

The calculator method is very straightforward, and is best understood by viewing Figure 8-14. In Figure 8-15, each box represents a key or combination of keys to be pressed on the calculator.

Using the distance and indicators discussed in the graphical method, and substituting the proper values into the calculator form yields the results shown in Figure 8-15. Note how the graphical solution and the calculator method yield the same results when rounded to the nearest mil.

Lines connecting boxes mean to copy the first box exactly into the next connected box. A plus (+) number in the final boxes means add shims; a minus (–) means to remove shims. Labels above the boxes indicate what data to put into that box. For horizontal alignment, a plus (+) means to move the machine to the left; a minus (–) means to move the machine to the right.

Once the correct amount of shims has been determined, miked and installed under the proper foot and hold-down torqued, a second set of readings should be taken, following the

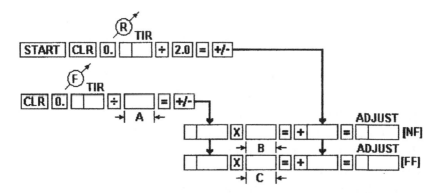

Figure 8-14. Rim & Face Calculator Method for Vertical Alignment

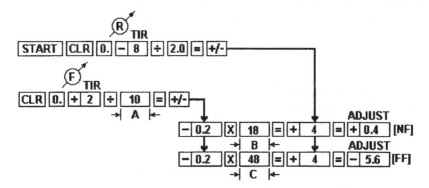

Figure 8-15. Rim & Face Vertical Alignment Solution

exact procedures as outlined in the first run. This will assure that the correct adjustments have been made. If the second run calls for additional adjustments, the correct amount of shims should once again be installed or removed. Certain machines, especially those with weak or semi-flexible frames, may require three or even four runs to obtain the exact alignment.

Remember, the amount of torque applied to the hold-down bolts, the sequence in which they are tightened, small burrs on the shims, dirt or grit can all influence the final outcome. By paying attention to the small details and being clean, an exact alignment can often be made on the first try. Once the fixtures are set up and the machine is down, and the alignment tools are on site, it takes very little time to make additional adjustments to obtain the exact desired alignment. This small effort will now pay dividends in extended equipment life.

Once the vertical alignment is completed, the horizontal alignment can be tackled. The primary difference in the horizontal alignment is that the rim indicator is set to zero at the 9 o'clock position, since bar sag plays no roll in horizontal plane. Zero both indicators at 9 o'clock and rotate the fixture and shafts to the 3 o'clock position to obtain the readings.

The layout of the graph and the calculator methods is performed exactly as in the vertical alignment. The only difference is that if the graph or the answer to the calculator method indicates a plus (+) reading, the foot needs to be moved to the right. If the reading or answer is negative, the foot needs to be moved to the left.

MOVING THE MACHINE
IN THE HORIZONTAL PLANE

Since this is a precision alignment method, banging the feet with a lead mallet is not the most desirable method to move the machine in the horizontal plane. However, there are several techniques that will perform well on most machines. Although all of these methods will not work on a given machine, one should be adaptable to most situations. With a little thought, you may devise other practical methods to precisely move the equipment.

1.   If the machine is equipped with horizontal bolts, tighten all jacking bolts snugly against the frame. Using the proper thickness feeler gauge, loosen the jack bolt on the side the machine needs to be moved toward, insert the feeler gauge and run the jack bolt against it, then remove the feeler gauge. Use the jack bolt on the opposite side to move the machine.

2.   Mount dial indicators on the feet of the machine, to measure the amount of movement at each foot location, on either side of the machine. Move the machine while observing the align-

ment dial indicators positioned in the 3 o'clock position. The face indicator should go to zero, while the rim indicator should go to 1/2 its reading.

3. Small machines can be moved using bar type glue clamps. A come-along and chain hoist can be employed for moving some large machines. With caution, jacking bars can be used to pry the machine into position. A small lead mall may be used to bump a machine into its final position.

Forms for the horizontal alignment are provided in the appendix.

Regardless of the method employed to physically move the machine, always assure the movement is measured and controlled. As with the vertical alignment, several adjustments may be required to achieve the final alignment. When the horizontal alignment is completed, check the vertical alignment again to assure it was not disturbed. Machines with uneven base plates could cause slight changes in the vertical alignment during the horizontal alignment process.

Remember, aligning a machine to within "tolerances" is not the goal of a good maintenance person. The more accurate the final alignment, the fewer coupling, seal and bearing failures.

THERMAL GROWTH AND HOT ALIGNMENTS

Many machines operate at either hotter or colder temperatures than when they are being aligned. Some refrigeration compressors operate at temperatures below freezing while some gas compressors may reach over 300°F. Changes in temperature cause these machines to either grow or shrink, and can distort the alignment. For these machines, consideration must be made for thermal growth.

Although most machines do not operate at these extremes, thermal growth should be considered for all alignment processes. Once the fundamental understanding of how to compensate for growth or shrinkage is understood, the few extra minutes required will assure the best possible final alignment.

Metals, like most materials, expand when heated. The amount they expand is expressed as the coefficient of thermal expansion. This coefficient is expressed as the change in length per degree of temperature rise. This is expressed mathematically as:

$$c = dL/dT \qquad\qquad (8.1)$$

By rearranging Equation (8.1) the amount of thermal growth for a given length (L) can be expressed as:

$$dL = c \times dT \times L \qquad\qquad (8.2)$$

Equation (8.2) states the change in length is equal to the difference in temperature times the coefficient of thermal expansion times the original length.

In general, most steels or cast iron frames have a coefficient of thermal expansion of about .0063 mils per inch per degree Fahrenheit, with aluminum and bronze being about twice that amount.

Table 8-3 lists the coefficient of expansion in mils per inch per degree Fahrenheit, for some common metals for use between 32°F and 212°F.

Table 8-3. Coefficients of Expansion

MATERIAL	COEFFICIENT OF EXPANSION
Soft Forged Iron	.0063
Cast Iron	.0059
Soft Rolled Steel	.0063
Hardened Steels	.0056
Nickel Steel	.0073
Aluminum	.0094
Bronze	.0100

Although these are not the exact coefficients of expansion for all steels and various alloys, they are close enough to provide solutions that will assure equipment is in exact alignment when it reaches operating temperatures.

Example 8-2

If a bar of soft rolled steel 12 inches long at 60°F is heated to 300°F, how long would the bar be?

Step 1. Using Equation (8-2) and finding the thermal growth coefficient from Table 8-3, the growth is .0063 × (300–60) × 12 = 18 mils. Therefore, the bar would be 12.018 inches long.

There are several ways to compensate for thermal growth:

1. Use the thermal offset provided by equipment manufacturer.
2. Calculate anticipated thermal growth at each foot.
3. Take a set of hot readings.

Of these methods, the third choice is by far the most desirable. It is accomplished by first aligning the equipment cold. That is, the machinery is cooled to ambient temperature. Prior to removing the alignment fixtures, their exact location is marked on the machine to assure they can be placed as close as possible to their original position. The machine is then placed on line and allowed to heat soak and stabilize at operating temperatures.

A sure sign of thermal growth problems is when a machine's vibration level increases with time after a cold start up. This is an indication that the unit is growing out of alignment.

After the machine has stabilized, it is shut down and the alignment fixture attached as soon as possible. A set of readings is taken in the same manner as with the cold alignment. Most equipment will require a considerable time to cool off, so there is adequate time to obtain a set of hot readings.

These hot readings can now be plotted over the original cold alignment graph, and the calculator method can be performed. If the machine experienced thermal growth or shrinkage, the results will indicate how many shims to add or remove. Of interest is the fact that when the machine is aligned in the future, it will need to be set exactly opposite, and by the same amount, as is indicated by these calculations. To do this precisely, the desired final indicator readings are required. That is, we do not want to set the machine at zero/zero when cold, but rather at some offset to compensate for the thermal growth.

If you are using the hot reading method, the final desired cold indicator readings are exactly opposite from the hot readings. This is best understood by studying the calculator method shown in Figure 8-16.

If you are working with known growth at the feet, either measured or provided by the manufacturer, then determining the desired readings from the graph is very straightforward:

R and F are the mils and direction measured from the final alignment line to the cold adjustable machine shaft centerline, at the rim and face indicator lines. Draw a line where the center of the adjustable machines shaft should be when cold. Next, simply count the direction and number of mils at each indicator line.

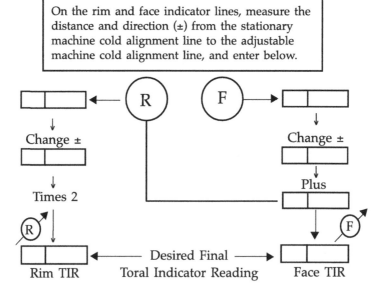

Figure 8-16. Determining the Final Desired Indicator Readings

An alternate method would be to measure the temperature of each supporting leg of both pieces of equipment, while the unit is operating at normal conditions. It is suggested that the temperature be taken every two inches from the base plate to the machine's shaft center line and the average temperature calculated for each foot location. This is due to the fact that most equipment supports are not of uniform thickness from their top to their bottom, and the fact that some heat is being dissipated along their length. These average temperatures along with the height from the base plate to the machine's centerline can be used to calculate the amount of anticipated thermal growth.

Even though this method is fairly accurate, it still is quite involved. Even using the manufacturer's offsets may not be as accurate as using the hot readings from the actual machine. Your equipment operates in a unique environment and under unique conditions. It's better to measure the actual growth rather than relying on the equipment manufacturer's estimated values.

Example 8-3

An electric motor is used to drive a chilled water pump. The pump was selected as the stationary piece of equipment. With the unit at normal operating conditions, temperature measurements were taken on the pump support legs and frame for thermal growth considerations. The ambient temperature during alignment was recorded at 78 degrees, and the unit had been down long enough to cool to room temperature. The following measurements were recorded:

Machine distances:
   A = 12"; B = 16"; C = 36"; D = 14"; E = 56";
   Pump Inboard leg height = 17";
   Pump Outboard Leg height = 22"

Pump average temperatures (running) previously recorded:
   Inboard = 106°F;
   Outboard = 138°F;

Vertical indicator readings:
    Rim = –4 mils;
    Face = 3 mils

What corrections should be made to assure proper alignment in the vertical direction with the pump at normal operating conditions?

Step 1. After a complete inspection for potential causes of unbalance, and determining the motor would be the adjustable piece of equipment, the equipment was checked for level and soft feet. The pump was also checked for soft feet and all shims were removed, cleaned and re-installed. After cleaning under all feet, the equipment was re-torqued and the first set of readings was taken. Both the graphical and the calculator methods were employed to assure the correct movement was achieved.

Step 2. Both readings have their algebraic signs changed and the rim reading is divided by two and plotted. A horizontal line is now drawn to the face indicator line and the face reading marked off from this point. The location of the face indicator reading is also circled. A line is now drawn from the two circled points on the indicator lines, through the lines representing the pump's feet. The distance to the cold alignment line is now measured.

In this example, the inboard foot shows a minus two (–2) and the outboard foot shows a minus seven (–7). If no thermal growth considerations were to be made, these readings would represent the amount of shim material to be added to each foot.

Step 3. Next, the amount of thermal growth (shrinkage) needs to be calculated. The pump case and support legs are constructed from cast iron with a coefficient thermal expansion of .0063 mills per inch per degree Fahrenheit. Using formula (8.2); Inboard: CL = .0063 × 17 × (106 – 78) = 2.99 mils; Outboard: CL = .0063 × 22 × (138 – 78) = 8.3 mils.

These answers are rounded to the nearest .001 inch or 1 mil,

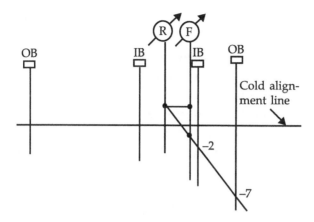

Figure 8-17. Cold Alignment Plot

based on practical shim thickness. The inboard would be thus be 3 mils and the outboard 8 mils. These two points are now marked off from current centerline and located on each of the inboard and outboard foot lines.

The two new points on the foot location lines are used to construct the hot alignment or final alignment line. This line represents where the centerline of the pump's shaft needs to be located when cold so that it will grow to the cold alignment line when it reaches operating temperature. By counting from these two points back to the cold alignment line, it is seen that both the inboard foot and the outboard foot need 1 mil of shim material removed.

Note that the same results could be obtained by adding the cold readings to the calculated thermal growth. For the inboard foot, the cold alignment correction was a minus two mils (–2), and the thermal growth was three mils (+3). Their sum is plus one mil (+1); thus, one mil of shim material is required. The outboard for cold correction was minus seven mils (–7), and the calculated thermal growth was plus eight (+8) mils. Their sum is plus one mil (+1), which is the same as measured on the graph.

Next, the actual indicator readings that will be present when the pump is misaligned for thermal growth need to be determined. This is accomplished by laying out the indicator positions in reverse.

By constructing the horizontal line from the point where the hot alignment line crosses the rim indicator line, both the rim and face readings can be measured. Simply count or measure the distance from the indicator locations back to the cold alignment line. In this example the rim reads plus one mil (+1) and the face reads zero. When the face reading is zero, the two shafts are parallel

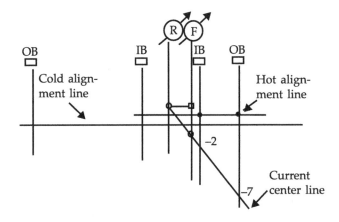

Figure 8-18. The Hot Alignment Line

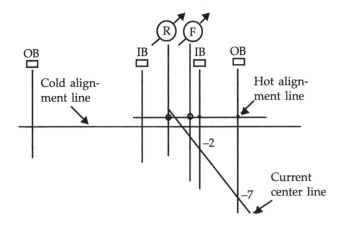

Figure 8-19. Determining the Indicator Readings

offset, and there is no angularity.

Continuing working backwards, the rim reading needs to be multiplied by two, and both readings need their algebraic signs changed. This results in a final set of indicator readings of plus two mils (+2) for the rim indicator and zero (0) for the face indicator. These readings should be recorded in the maintenance file for future reference.

Taking the keystrokes from Figure 8-14 and putting them into equation form yields:

$$IB = -1 * \left(\frac{FB}{A} + \frac{R}{2} \right) \tag{8.3}$$

$$OB = -1 * \left(\frac{FC}{A} + \frac{R}{2} \right) \tag{8.4}$$

Where:

F = Face indicator reading.
R = Rim indicator reading.
A = Diameter traced by the face indicator.
B = Distance from the rim indicator stem to the center of the inboard hold-down bolt.
C = Distance from the rim indicator stem to the center of the outboard hold-down bolt.

Substituting into these formulas, the results for Example 8-3 yield minus two mils (–2) for the inboard foot and minus seven mils (–7) for the outboard foot. Once again the calculated thermal growth is simply added to these readings to obtain the final alignment line.

Using Figure 8-16 and converting the key strokes into equations yield:

$$R_f = -R \times 2 \tag{8.5}$$

$$F_f = -F + R \tag{8.6}$$

Where:

R_f = Final desired rim indicator reading.
F_f = Final desired face indicator reading.

Substituting the readings from Example 8-3 into Equations (8.5) and (8.6) yields plus two mils (+2) for the rim indicator and zero (0) for the face indicator.

Either method can be employed to achieve good results. In general, the calculator method is slightly more accurate. The larger the scale chosen for the graph, the better the accuracy. However, it is always advisable to work both methods to assure there is no mistake.

When approaching an alignment problem, break it down into simple parts. If an equipment train consisted of a steam turbine, gearbox, centrifugal pump, a second gearbox, and a diesel engine, the problem is no more complex than a simple pump and electric motor. Establish the stationary piece of equipment, and then align the next piece to it. It then becomes the stationary piece of equipment for the subsequent alignment. Continue until all the equipment is aligned. This will work even if the chosen stationary equipment is in the center of the train of equipment.

EQUIPMENT WITH DRIVESHAFTS

Dealing with equipment with driveshafts, such as aligning a cooling tower fan, is also a straightforward process. The fixture is first set up across the coupling from the motor to the drive shaft, and readings are taken. Next, the fixture is moved and placed on the drive shaft to read across the coupling to the gearbox. Again readings are taken.

In the first case, the B and C distances were from the gear box feet to the rim indicator stem, while it was mounted to the motor shaft. In the second case, the B1 and C1 distances are from the gear box feet to the new indicator location. By superimposing these two graphs, the total correction required is shown. Simply count or measure the corrections for each foot and add them together. The result is the actual correction required. Thus, you only need to make one corrective move for both couplings.

In Figure 8-20, if the first run inboard correction is plus five

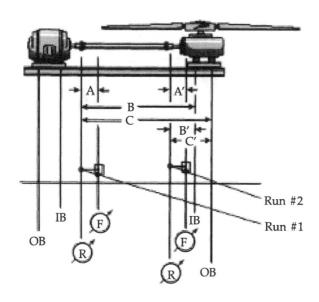

Figure 8-20. Aligning Equipment with Driveshafts

mils (+5) and the second run inboard correction is minus five mils (–5) the result is zero. Thus the inboard foot is ok. If the first run outboard correction was plus nine mils (+9) and the second run outboard corrections is minus one mil (–1), the result is to add eight mils (+8) to the outboard feet.

U-JOINT COUPLINGS

One word of caution: Some companies have changed to U-joints in the cooling tower fan drive shafts. Often, maintenance personnel attribute premature failure of these U-joints to corrosion, because after one fails, the needles and cup are filled with rust. This may be due to improper alignment. A U-joint must be misaligned a proper amount to assure proper lubrication of the needle bearings. Too much misalignment will cause large velocity fluctuations and uses excessive horsepower; too little, and the U-joint bearings will fail due to a lack of lubrication.

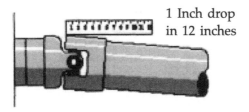

1 Inch drop
in 12 inches

Figure 8-21. Proper Alignment of a U-Joint Coupling

The proper misalignment for most applications is 4 to 6 degrees. Using the mid-point value of 5 degrees, the drive shaft would need to rise or drop 1.04 inches per foot of length. This should be rounded off to one inch per foot of drive shaft length. This alone does not excuse the use of precision alignment, since the misalignment need be only in a single plane. This means that the horizontal alignment should still read zero/zero.

SPOOL PIECES

Frequently, equipment trains are coupled using a spool piece between the two pieces of equipment.

Spool pieces can be measured across just as though they were a coupling, provided you can build a large enough indicator bracket. Since the bar sag is measured and compensated for, the amount of droop in the fixture is of no concern.

In the case where the spool piece is too long, it is aligned in the same manner as equipment with a driveshaft. That is, two sets of readings are taken and their sum is used to adjust the alignment. See Figure 8-20 for details.

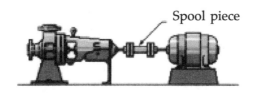

Spool piece

Figure 8-22. Equipment with Spool Pieces

BORE READINGS

Some equipment will need to be aligned using a bore reading rather than a rim reading. This is found in many large gas turbine/compressor applications. A bore reading is simply an upside-down rim reading and can be handled as such in the alignment process. All that is required is to set the positive (+) bar sag at 6 o'clock and take the reading at 12 o'clock.

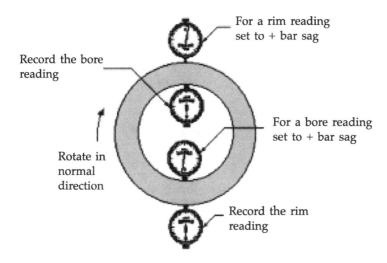

Figure 8-23. Bore Readings

VERTICAL PUMPS

In the case of vertical pumps, the first step is to remove all old shims, clean the mounting surfaces, and then correct any soft feet. The angularity is corrected first, and then the parallel offset.

Correcting the angularity is easily accomplished with the face indicator when using the rim and face method of alignment. The pump is viewed from above and labeled as shown in Figure 8-25.

Figure 8-24. Vertical Pump

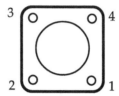

Figure 8-25. Top View of Vertical Pump Mount

Label the hold-down bolt locations and measure the distances between them. Set the dial indicator to zero at any of the locations. Rotate the indicator in the direction of normal operation and record the readings at the other hold down locations. Find the largest positive reading. **Note:** If the highest point was selected as the start point, zero will be the highest positive reading.

The readings are now adjusted to the highest positive reading by subtracting the highest positive reading from all readings. The process is best understood through an example.

Example 8-4

A vertical pump has four hold-down bolt locations which are labeled as in Figure 8-25. The distances between hold-down bolts are equal and found to be 14 inches. The diameter traced by the face indicator is set at 12 inches and the indicator is zeroed at point 1. The following readings were recorded: 1 (or 0 degrees): indicator set to 0 mils; 2 (or 90 degrees): –1 mils; 3 (or 180 degrees): +2 mils; 4 (or 270 degrees): –2 mils. What corrections are required?

Step 1. The readings are now adjusted by selecting the largest positive reading as zero, and subtracting it from all readings, as follows: 1 (or 0 degrees): 0 mils – 2 mils = –2 mils; 2 (or 90 degrees): –1 mils – 2 mils = –3 mils; 3 (or 180 degrees): +2 mils – 2 mils = 0 mils; 4 (or 270 degrees): –2 mils – 2 mils = –4 mils.

As seen from the above, point 3 was the highest hold-down location and the others will be adjusted to the same plane as point 3.

Step 2. Next, the slope is calculated. That is the change over the diameter A traced by the face indicator. This is used to correct the readings back to the hold-down locations.

Referring to Figure 8-26, it can be seen that the distances between the four locations where readings are recorded are exactly A/2 apart in the X and Y directions.

After the readings were corrected, point 3 became the zero reference point. The slope for the reading from point 3 to point 4 (X direction) is: –4/(A/2) = –4/6 or –.667 mils per inch. Likewise, the slope from 3 to 2 (Y direction) is thus –3/(A/2) = –3/6 or –.500 mils per inch.

Since the hold-down location 3 is the reference point, all that remains is to calculate the correction amount for the other locations. Point 4: 18 * –.667 = –12 mils; point 2: 18 * – .50 = –9 mils; point 1: is the algebraic sum of points 2 and 4 or –21 mils.

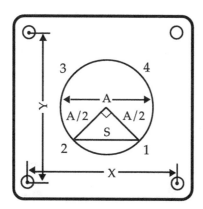

Figure 8-26. Correcting the Readings

Since the rim readings are all in the horizontal plane, no compensation for bar sag is required. Zero the indicator on one surface, such as the 1-2 side and take a reading on the opposite side, 180 degrees away. Set the dial indicator to 1/2 its reading and move the motor until the indicator reads zero. Repeat this process for the side 90 degrees from the initial side, the 2-3 side.

Check the final alignment by repeating the readings to assure the motor is centered over the pump.

THE CATENARY

Some long shafts may exhibit deflection due to their own weight, or in combination with the weight of a mounted component that can affect the alignment process. Perhaps the best way to demonstrate the potential effect is through examples.

Example 8-5
A large roller is made of steel and is 6 inches in diameter. It is supported by two bearings that are 12 feet apart. How much will the center of the shaft sag? How much angular deflection will be present at the end of the shaft?

Step 1. Steel weighs approximately 480 pounds per cubic foot, so the weight of the portion of the shaft between the bearings is its volume times the density $W = V \times d$ or $W = L \times A \times d = (12 \times \pi \times (.5)^2/4) \times 480$ or approximately 1,131 pounds.

Step 2. The deflection for a solid shaft is:

$$dx = 5wL^3/384EI \hspace{3cm} (8.7)$$

Figure 8-25. A Shaft Deflecting Under its Own Weight

The moment of inertia (I) for a solid shaft is:

$$I = \pi d^4 / 64 \qquad\qquad (8.8)$$

If the modulus of elasticity (E) for steel is assumed to be 30 $\times 10^6$, and

I = 63.62. The total deflection at the center is dx = 5(1131)(144)^3/384(30 × 10^6)(63.62) = .02304 inches.

Step 2. Since the catenary for shafts deflecting due to their own weight is very flat, the tangent at the bearing location is closely approximated by a simple triangle. That is, the curvature of the catenary approaches a straight line from its center to the bearing location. This slight error will have no effect on the final answer.

The slope of the curve at the bearing is dy/dx or .02304/72 = .00032. Since the tangent is also defined as the slope at any point, the arctangent function provides the angle the face of the shaft makes with the vertical plane. Thus arctangent of .00032 is .0183 degrees.

Step 3. The offset in the face reading due to the deflection of the shaft is then expressed as dF = d sin(A). Since the diameter (d) is 6 inches, dF = 6 × sin(.0183) = .00192 inches or 1.9 mils. This would be rounded up to 2 mil offset for the face reading. From this ex-

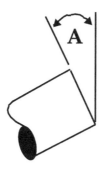

Figure 8-28. The Angle of the Face Due to Deflection of a Shaft

ample, it can easily be seen that any alignment process that did not account for the sag of large or heavy shafts would result in a poor alignment.

As a side note, good design practice allows for a maximum shaft deflection of .01 inches per foot of shaft length. In this example, the deflection per foot of shaft length is .02304/12 = 0.00192 inches per foot—well within design limits.

Example 8-6

A centrifugal pump has a 2-inch diameter shaft that is mounted between two bearings 60 inches apart. The rotor element weighs 448 pounds and is mounted 20 inches from the left-hand bearing. Assuming the rotor has a uniform weight distribution, what is the total deflection at the center of the shaft? What is the face offset required for proper alignment?

Step 1. First the deflection of the shaft due to its own weight is calculated in the same manner as in Example 8-4. The weight of the shaft is W = 52.36 pounds. The moment of inertia is .7854, and the maximum deflection is found to be 0.00625 inches. These answers could easily be looked up in the tables provided in the appendix section.

Step 2. Now the weight of the rotor must be accounted for. The sum of the deflection due to the shaft weight and the deflection due to the concentrated load are added together to obtain the total deflection.

For a non-central load:

$$dx = [W * (c/3)/(E * I * x')] * \{[c * (x'/3) + c * (c'/3)]^3\}^{1/2}$$
(8.9)

Where:

   c   = distance from left-hand bearing to the concentrated load.

   c' = distance from the right-hand bearing to the concentrated load.

$$x' = c + c'$$

In this example, $x' = 60$; $c = 20$; thus c' must equal 40. Since I has already been calculated and the weight (W) is given, $dx =$ [448 * (20/3)/(30000000 * .7854 * 60)] * {[(20 * (60/3) + 20 * (40/3)]^3}$^{.5}$ = 0.03636 inches or 36.36 mils.

The total deflection is 0.00625 + 0.03636 or 0.042615 inches. Since half the length of the shaft is 30 inches, the slope at the bearing is 0.042615/30 or 0.00142. Taking the arctangent of this slope yields an offset angle of 0.08139 degrees. The sine of this angle is found to be 0.001420.

Step 3. Multiplying the sine times the diameter gives the face offset or (2) * .001420 = 0.0028410 inches or 2.8 mils. In this case, the total deflection per foot of shaft length is 0.0085 inches and is just within design limits.

This deflection of the stationary shaft is treated the same way as thermal growth. Since the face offset is 2.8 mils over 2 inches, this same slope is used to locate the cold alignment line. A line is drawn from the rim indicator line with a slope of 1.4 mils per inch. The horizontal line is now the final alignment line for the center of the adjustable machine. This is best illustrated through an example.

Example 8-7

The machine in Example 8-6 is determined to be the stationary machine. It was found to have a catenary effect of 1.4 mils per inch. The adjustable machine has a B distance of 18 inches and a C distance of 48 inches. The A distance for the face indicator is 10 inches. A rim reading of +10 mils and a face reading of –5 mils was obtained. What corrections are required?

Step 1. First, a new machine measurement is required, that is the distance from the stem of the rim indicator to the first bearing of the stationary machine. This measurement is labeled G, and found to be 12 inches.

As shown in Figure 8-29 the cold alignment line is drawn from the first bearing location of the stationary machine with a slope of 1.4 mils per inch. This line represents the current position of the centerline of the stationary machine shaft.

All indicator readings will be measured from this newly constructed cold alignment line. This will assure, once the machine is

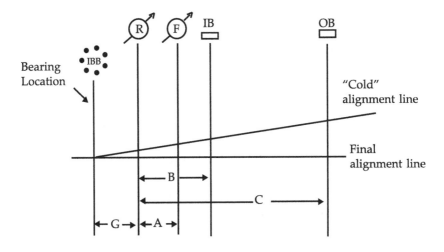

Figure 8-29. Example 8-7 Machine Measurement Layout

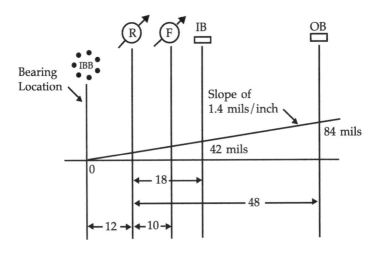

Figure 8-30. Centerline of the Stationary Machine's Shaft

operational and the catenary effect removed, the two shafts will be in alignment.

As the stationary machine becomes operational, its centerline will align with the final alignment line. However, while it is at rest, the adjustable machine must be referenced to its cold, or rest position.

Step 2. Next, the rim reading is divided by two and has its algebraic sign changed. It is then marked off from the current position of the stationary machine's shaft centerline. A horizontal line is drawn from this point to the line representing the stem of the face indicator. The face reading is now laid off from this new point. The two readings' points are now connected with a line that extends beyond the adjustable machine's outboard foot line.

The distance from this new line to the final alignment line is now measured at both foot locations for the adjustable machine. The magnitude and direction are recorded and found to be plus 3 mils (+3) for the inboard foot and minus 12 (–12) mils for the outboard foot.

As with all alignment graphs, the positive reading requires shims to be removed and the negative reading requires shims to be added. In this example, 3 mils of shims would be removed from the inboard foot and 12 mils of shims would be added to the outboard foot.

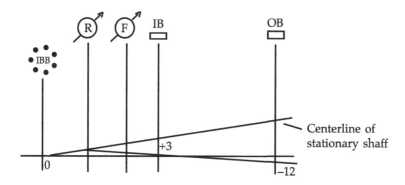

Figure 8-31. Final Alignment Requirements

50 STEPS TO ACHIEVE PRECISION ALIGNMENT

1. Assure the equipment is locked out and tagged out.
2. Determine which machine will remain stationary.
3. Remove the coupling guard and other obstructions.
4. Inspect coupling bolts, nuts, washers and keys, and repair as required.
5. Remove all old shims from stationary machine and clean under feet. Discard any old shims that have burrs, are torn, or are made of materials other than stainless steel, brass or plastic.
6. Check stationary equipment for soft feet and repair as required.
7. Level, center and remove any piping or other stresses from the stationary equipment, if conditions warrant.
8. Use no more than three shims under any foot.
9. Torque all hold-down bolts.
10. Check coupling end float by barring both machines in the axial direction. Correct if necessary.
11. Remove all shims from the adjustable machine and clean under its feet.
12. Check adjustable machine for soft feet, and correct if necessary.
13. Assemble the alignment fixture on the equipment and mark its location.
14. Measure and record all machine dimensions.
15. Remove fixture and measure and record bar sag.
16. Lay out the basic graph dimensions.
17. Determine if thermal growth will be considered.
18. Re-install fixture on machine, using marks as a guide to the correct location.
19. Zero the face indicator and set the rim indicator to the plus bar sag reading at the 12 o'clock position.
20. Rotate equipment to the 6 o'clock position in the normal direction of operation using caution not to go past the mark.

21. Observe the indicators as the shafts are rotated, to determine the direction they move.
22. Take and record readings at the 6 o'clock position.
23. Continue to rotate in the proper direction and verify the 12 o'clock readings.
24. Plot the readings on the graph.
25. Use a calculator to verify the required machine movement.
26. Adjust the machine as required in the vertical plane.
27. Repeat steps 19 through 26 as required.
28. Place the indicators on the left side of the machine, and zero the indicators.
29. Mark the left side of the machine.
30. Rotate the machine 180 degrees to the right side.
31. Observe the indicators as the machine is rotated.
32. Take and record the indicator readings.
33. Rotate the machine in the proper direction and verify the left side zero.
34. Plot the readings on the graph.
35. Verify machine movements with the calculator method.
36. Use dial indicators or feeler gauges to measure machine movement.
37. Use indicators on fixture if necessary, by setting both to 1/2 their readings and moving the machine until both read zero.
38. Repeat steps 28 through 36 as required.
39. Verify both the vertical and horizontal alignments.
40. Record all soft feet corrections, all shims used, and any offset for thermal growth.
41. Record the temperature of the equipment at the time of alignment.
42. Assemble the coupling guard and other required parts and operate the equipment until it is heat soaked.
43. Shut down equipment and lock and tag out.
44. Remove coupling guard and other necessary parts and reinstall the fixtures in their original position, using the marks on the shaft.

45. Take vertical and horizontal readings as detailed above.
46. Plot the results on the original graph, to determine thermal growth.
47. Make any required adjustments to the machine.
48. Record all final data.
49. Reassemble and run equipment.
50. Measure and record the final vibration levels.

The above 50 steps pertain only to the advanced rim and face alignment process. In the following chapter, the reverse indicator method of precision alignment is presented.

Chapter 9

Reverse
Indicator Alignment

INTRODUCTION

I n the previous chapter, an advanced rim and face alignment
process was examined in detail. In this chapter, a double rim
method known as the reverse indicator method will be discussed. The alignment processes will be found to be quite similar,
and there will be machines that will be more easily adapted to one
or the other methods. Both methods are capable of accuracy of less
than 1 mil.

As with any alignment process, several preliminary steps
must be performed. The stationary machine must be identified,
and the left side of the equipment designated. Next, the machine
is inspected for level, soft feet, damaged or excessive shims and
for sources of unbalance. Follow the process outlined in the rim
and face method to accomplish these inspections.

The basic difference in these two methods is the types of
readings taken. In the rim and face method, the readings were
both taken from the stationary machine to the adjustable machine.
In the reverse indicator method, one reading is taken from the
stationary machine to the adjustable machine, while the other is
taken from the adjustable machine to the stationary machine. This
is necessary, since both readings are rim readings.

As with the rim and face method, disassembly of the coupling is not necessary, but a close inspection of the coupling and
its components is highly recommended. Once all inspections are
complete, the indicator fixtures should be assembled on the machine.

Once the indicator fixtures are properly assembled on the machine, and a trial set of readings taken to assure the fixture clears the structure when rotated a full 360 degrees, its location on the two shafts should be clearly marked. The fixtures are now removed from the machine and placed on a ridged pipe or shaft to determine bar sag. Note that in this method, both indicators read a rim value, and thus both indicator fixtures must have their bar sag accounted for.

BAR SAG

Once again, determining the bar sag is a straightforward process. The indicators are set to zero in the 6 o'clock position and rotated to the 12 o'clock position where the bar sag is read. Be sure to determine the proper bar sag for each indicator fixture. Even though the two fixtures are assembled the same, they may have different amounts of sag.

It is doubly important to account for bar sag using this method of alignment, since both indicator readings are affected.

The indicator fixture is thus assembled so that one indicator

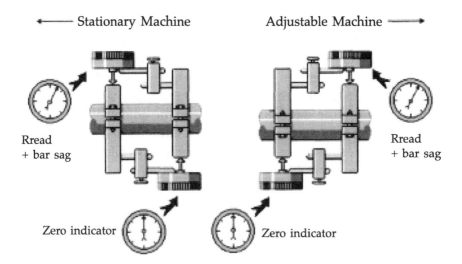

Figure 9-1. Measuring Reverse Indicator Bar Sag

is placed on each machine. Figure 9-1 shows how the bar sag is determined for reverse indicator method.

Once the bar sag values have been determined and recorded, assure that both indicators are set to the plus bar sag when they are in the 12 o'clock position. No further action is required to compensate for bar sag.

The fixtures can now be reassembled to the machine, using the reference marks as guides. An initial set of readings can now be taken to assure the fixtures are properly assembled and sufficiently ridged to return the indicators to their plus bar sag readings at 12 o'clock, after being rotated 360 degrees.

MACHINE MEASUREMENTS

Now the machine distances can be measured and recorded. As with the rim and face method, the A, B, and C distances are required, and the D and E values if thermal growth is to be considered. Again the A distance is the most critical and should be measured to within 1/8". All distances are measured from the stem of the stationary machine indicator to the respective point. Note that in this method, the A distance is measured to the stem of the adjustable indicator. Figure 9-2 shows the machine distances for this method.

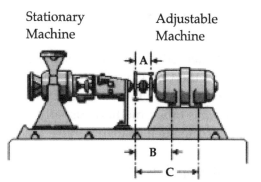

NOTE: "A" is the distance between the two indicator stems.

Figure 9-2. Reverse Indicator Basic Machine Measurements

VERTICAL ALIGNMENT

As with any precision alignment method, the vertical alignment should be corrected first. Again, use caution to rotate the equipment in its normal operation direction, and never back up to obtain a reading.

Starting with one of the dial indicators at the 12 o'clock position, and set to its plus (+) bar sag, rotate the machine in its normal direction of operation to the 6 o'clock position. Be careful to observe the movement of the indicator as it passes through the 180 degrees of rotation, noting its direction of travel. The indicator is now read and recorded. Note that the other indicator is now in the 12 o'clock position and should be set to its plus (+) bar sag.

Rotate the machine 180 degrees. Bring the first indicator to the 12 o'clock position where it should return to its plus (+) bar sag reading. Bring the other to the 6 o'clock position where the second reading is taken and recorded. Rotate the machine 180 degrees again, and assure the second indicator returns to its plus (+) bar sag reading at 12 o'clock.

The data can now be transferred to the graph, used in the calculator method, or input to a computer program.

GRAPHICAL SOLUTION

The graphical method employs the same principles used in the rim and face method of Chapter 8. A suitable scale for the vertical and horizontal scales must be determined. Use the same principles used in the other method, usually 1 mil per inch in the vertical, and a scale such that the entire machine will fit the graph.

Remember, all graphical points are measured from the stem of the stationary machine dial indicator. Starting at the left of the graph, draw the first vertical line and properly label it the outboard foot of the stationary machine. Use the scale factor and the E distance to locate the stationary machine dial indicator line. Draw and label this line. All other points are now measured from this line. Figure 9-3 shows the final graph layout.

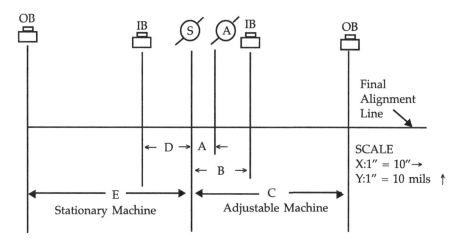

Figure 9-3. Reverse Indicator Graph Layout

Note the similarity of this graph layout with the rim and face graph. The only two differences are the labeling of the indicator lines and the way the A distance is measured on the machine.

To plot the data, the stationary indicator reading is divided by two (remember rim readings are always divided by two in any method), and then laid off on the stationary indicator line. Remember positive numbers are counted up from the cold alignment line, and negative numbers down. Next, the adjustable indicator reading is divided by two, and then its algebraic sign is changed. Again, the sign is changed to transfer the reading back to the stationary machine. This number is now laid off on the adjustable indicator line.

A straightedge is used to draw a line connecting these two points, and extending completely past the adjustable machine's outboard foot line. Counting from the intersection of the foot lines to the cold alignment line determines the amount of required movement. Again in the vertical alignment, if the line is above the cold alignment line, shims must be removed. If the line is below the cold alignment line, shims must be added.

Example 9-1

A machine is measured and the following data are recorded: A = 10, B =18, C = 48, the stationary indicator STIR = + 8, and the adjustable indicator ATIR = - 4. What adjustments are required to properly align the adjustable machine to the stationary machine?

Step 1. First, the stationary reading is divided by two, yielding a plus four (+4) which is laid off four divisions above the cold alignment line, along the stationary indicator line. Note that a scale of one division per mil was chosen for the vertical axis.

Step 2. Next, the adjustable indicator reading is divided by two and its algebraic sign changed, yielding a plus two (+2). This value is laid off two divisions above the cold alignment line, along the adjustable indicator line. Figure 9-4 shows the completed graph for the example.

Refer to Figure 8-13 to see this same machine aligned using the rim and face method. Even though the indicator readings were different, the answer comes out the same. This illustrates that either alignment method will yield the same results. The larger the

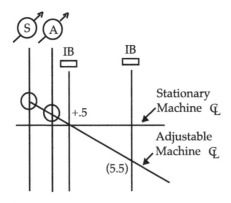

Figure 9-4. Reverse Indicator Example Plot

scale and the more time taken to assure the drawn lines intersect the indicator points at their centers, the more accurate the results.

CALCULATOR METHOD

As with the rim and face method, there is a simplified calculator method. Figure 9-5 shows the key strokes for this solution.

If the same machine measurements are used, and the indicator readings from the reverse indicator example are input, Figure 9-6 shows the results.

The calculation method indicates 0.4 mils to be removed from the inboard foot and 5.6 mils to be added to the outboard foot. Both methods indicate the same basic required changes.

The calculator method is slightly more accurate in this case. If a larger scale were used in the graphical method, the exact same results would occur. However, adding or subtracting less than 1 mil is not practical, and in this example, 6 mils would be added to the outboard foot and no change would be made to the inboard foot.

Viewing the calculator method in equation form:

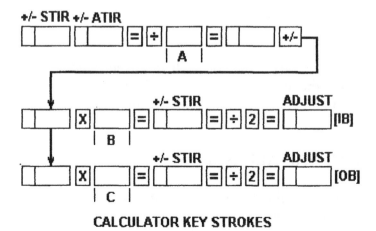

CALCULATOR KEY STROKES

Figure 9-5. Reverse Indicator Calculator Key Strokes

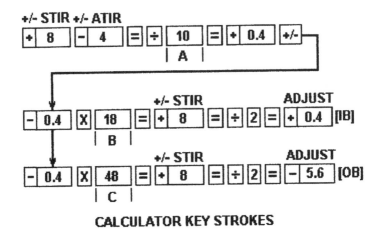

CALCULATOR KEY STROKES

Figure 9-6. Reverse Indicator Calculator Results

$$IB = \frac{\left[\left(\dfrac{STIR + ATIR}{A}\right) * B\right]}{2} - STIR \tag{9.1}$$

$$OB = \frac{\left[\left(\dfrac{STIR + ATIR}{A}\right) * C\right]}{2} - STIR \tag{9.2}$$

Where:

STIR = the total indicator reading for the stationary machine.

ATIR = the total indicator reading for the adjustable machine.

A, B and C are the machine measurements

HORIZONTAL ALIGNMENT

Once again, the horizontal alignment is performed exactly the same as the vertical alignment, with the exception that both indicators are set to zero on the left side or 9 o'clock position. Use the same cautions and steps outlined for the vertical alignment to graph the results and calculate the solution. Again the calculator should provide an exact solution, while the graph should be within less than 1 mil of the calculator.

The moving of the machine during the horizontal alignment was discussed in the rim and face chapter, and should be reviewed. Since this method employs two rim indicators, they can be set to 1/2 their readings and observed as the machine is moved. Note however that both indicators should be set to 1/2 their value while in the 3 o'clock position only. This means that when the indicator is positioned at 9 o'clock, it will read minus 1/2 its 3 o'clock reading. With this, the machine can be moved until both indicators read zero. This method can be used on small machines or machines that are easy to move in a controlled manner. However, measuring the movement of the machine at each foot is the recommended method, and will prove more accurate in most applications.

The best solution is to mount dial indicators to the base so that they read against the feet. Set both indicators to zero and use a controlled method of moving the machine. Move the machine until the required readings are observed on the dial indicators. A final set of both vertical and horizontal readings must be taken to assure the vertical alignment was not upset and that the correct adjustment was made in the horizontal direction.

THERMAL GROWTH

Thermal growth is basically handled the same as it was for the rim and face method. Again, the most desirable method is hot readings obtained from the machine after it has been aligned cold. Use the same procedures to obtain a set of hot readings, and plot their values over the cold graph. Note if a calculator solution is desired, calculate the results of the hot run as with any other run, and simply add the two runs together algebraically to obtain the final results.

Remember, it is the objective to misalign the machine when cold to assure proper alignment under normal operating conditions. In any alignment process, this is accomplished by aligning the machine opposite the hot readings.

When using this method to offset the equipment to compensate for thermal growth, be sure to record the temperature of all equipment during the alignment process. This will assure the most accurate final alignment. After any alignment process, be sure to measure and record the final vibration amplitudes and record them on the alignment form for future reference.

Example 9-2

The machine in Example 9-1 has known thermal growth for both the stationary machine and the adjustable machine. The outboard foot of the stationary machine grows vertically 3.5 mils and the inboard foot grows 2 mils. The D distance is measured at 12 inches and the E distance at 56 inches. The adjustable machine will grow 1 mil at the inboard foot and shrink 2 mils at its outboard foot. Determine what corrections need to be made to perfectly misalign the machines to assure proper alignment when at operating temperatures. What should be the final indicator readings?

Step 1. The first step is to lay out a graph of the equipment and determine the thermal growth effects. The hot centerlines for both machines will be plotted to determine their final positions.

The basic graph layout is illustrated in Figure 9-7. Note again that all distances are measured from the stem of the stationary indicator.

Step 2. Next, the hot position of each foot is located and the centerlines for the hot alignment are drawn.

In this example, the stationary machine hot centerline crosses the inboard foot of the adjustable machine at plus 1 (+1) mil and the outboard foot at zero. The hot centerline of the adjustable machine also is at plus one mil at the inboard foot and minus two (–2) mils at the outboard foot. Both machines will grow to the same point on the adjustable machine inboard foot, and thus no correction is required. Since the adjustable machine hot centerline

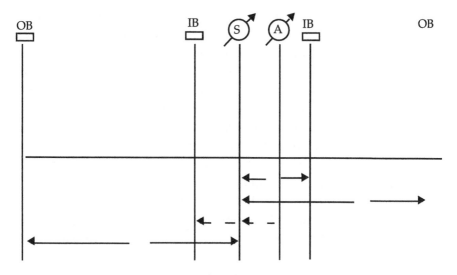

Figure 9-7. Example 9-2—Basic Layout

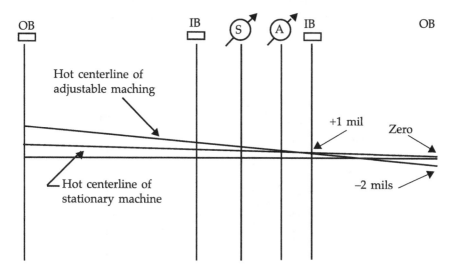

Figure 9-8. Example 9-2 - Hot Centerlines

on the outboard foot line is 2 mils below the stationary machine hot centerline, the final alignment will place the adjustable machine outboard foot high by two mils.

Since the cold alignment was worked out in Example 9-1 and found to require no change on the inboard foot and that 6 mils were required to be added to the outboard foot, the two results are simply added together. In this example no change to the inboard foot was required from either calculation.

Therefore no change will be made to the inboard foot. The outboard foot required 6 mils of shims to be added for the cold alignment and required to be set an additional 2 mils high to compensate for the hot alignment. Adding these two together requires a total of eight (+8) mils to be added to the outboard foot.

Step 3. Next, a new plot of the final desired cold alignment is made. This is shown in Figure 9-9.

Now all that is required is to determine the final indicator readings, when the machines are perfectly misaligned cold.

Figure 9-10 illustrates how to calculate the final desired indicator reading to assure proper misalignment, when the machine is cold. In this example, the cold alignment line crosses the station-

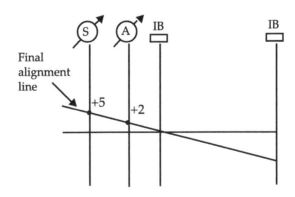

Figure 9-9. Example 9-3 - Final Cold Alignment Line

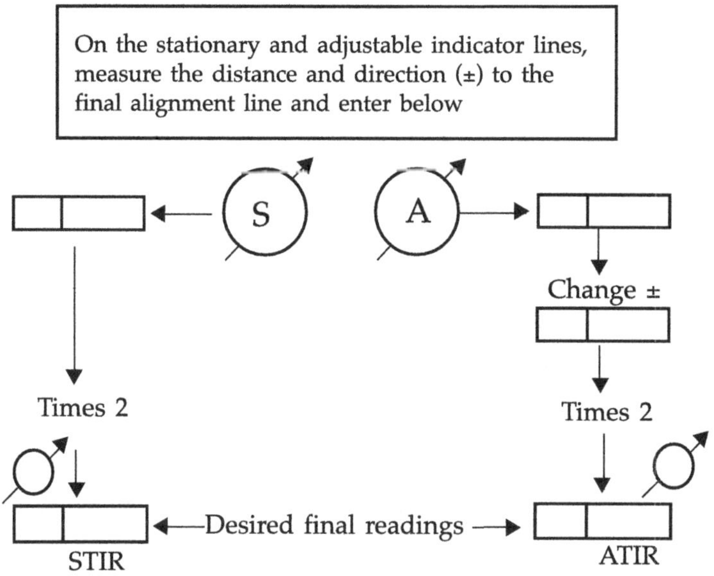

On the stationary and adjustable indicator lines, measure the distance and direction (±) to the final alignment line and enter below

S   A

Change ±

Times 2   Times 2

STIR   ←—Desired final readings —→   ATIR

Figure 9-10. Final Alignment Readings for Reverse Indicator Alignment

ary indicator line at plus five (+5). Therefore the stationary indicator reading (STIR) should be plus ten (+10). The cold alignment line crosses the adjustable indicator line at plus two (+2). The final adjustable indicator reading (ATIR) should be minus four (–4).

One final word about alignment. Regardless of the method employed, when readings are taken from one machine to the other, the readings always must have their algebraic signs changed, so that the readings are transferred back to the original machine.

In the reverse indicator method, the stationary machine reading was not changed, since it was already a reading on the stationary machine. Note however that all other readings presented in either method required transferring back to the stationary machine, since they were read across the coupling.

Also, all rim or bore readings must be divided by two, regardless of the alignment technique employed.

BORE READINGS

Some machines require bore readings to be taken for the alignment process. This could be a bore and face, a bore and rim, or a bore and bore alignment process. Any of the three systems mentioned above is easily handled with the methods previously outlined. Remember, a bore reading is nothing more than an up-side-down RIM reading. Thus all that is required is to treat every process involving a bore reading exactly opposite of a rim reading.

The bar sag is set at 12 o'clock as a negative number, since the indicator is mounted upside-down. Bore readings also must be divided by two; however, their algebraic sign should not be changed when transferring across a coupling. The remainder of the alignment process is exactly the same as dealing with rim readings.

Torque Specifications

Torque Values in Ft.-Lbs.

Bolt Dia.	Threads per inch	Grade					
		2 *Dry*	2 *Oiled*	5 *Dry*	5 *Oiled*	8 *Dry*	8 *Oiled*
1/2	13	38	31	75	55	110	80
1/2	20	52	42	90	65	120	90
9/16	12	52	42	110	80	150	110
9/16	18	71	57	120	90	170	130
5/8	11	98	78	150	110	220	170
5/8	18	115	93	180	130	240	180
3/4	10	157	120	260	200	380	280
3/4	16	180	130	300	220	420	320
7/8	9	210	160	430	320	600	460
7/8	14	230	175	470	360	660	500
1	8	320	240	640	480	900	680
1	12	350	265	710	530	990	740
1-1/8	7	480	385	795	580	1430	1070
1-1/4	7	675	540	1105	805	1975	1480
1-3/8	6	900	720	1500	1095	2650	1980
1-1/2	6	1100	880	1775	1295	3200	2400
1-5/8	5.5	1470	1175	2425	1770	4400	3300
1-3/4	5	1900	1520	3150	2300	5650	4235
1-7/8	5	2360	1890	4200	3065	7600	5700
2	4.5	2750	2200	4550	3320	8200	6150

The above table is provided for reference only. Always consult the manufacturer's recommended torque specifications if available. When tightening fasteners, it is recommended that the final torque value is reached in three stages—that is, first torque to 33%, then 66%, and then the final value. In addition, assure that a proper tightening pattern is used.

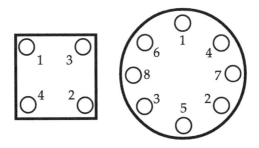

Figure A-1. Some Basic Torque Sequences

Some newer machines may have metric bolts and the torque specifications may be given in Newton-Meters. To convert Newton-Meters into foot-pounds of torque:

1 Meter = 3.28333 Feet
1 Newton = .225 Pounds
1 Foot x 1 Pound
_____

.30456 Meters x 4.4444 Newtons
Foot-Pounds = Newton-Meters / 1.3536   (A-1)

Trigonometric Tables

TRIGONOMETRIC TABLES (0-30)

ANGLE	SIN	COS	TAN
0	**0.00000**	**1.00000**	**0.00000**
1	0.01745	0.99985	0.01746
2	0.03490	0.99939	0.03492
3	0.05234	0.99863	0.05241
4	0.06976	0.99756	0.06993
5	**0.08716**	**0.99619**	**0.08749**
6	0.10453	0.99452	0.10510
7	0.12187	0.99255	0.12278
8	0.13917	0.99027	0.14054
9	0.15643	0.98769	0.15838
10	**0.17365**	**0.98481**	**0.17633**
11	0.19081	0.98163	0.19438
12	0.20791	0.97815	0.21256
13	0.22495	0.97437	0.23087
14	0.24192	0.97030	0.24932
15	**0.25882**	**0.96593**	**0.26795**
16	0.27564	0.96126	0.28675
17	0.29237	0.95630	0.30573
18	0.30902	0.95106	0.32492
19	0.32557	0.94552	0.34433
20	**0.34202**	**0.93969**	**0.36397**
21	0.35837	0.93358	0.38386
22	0.37461	0.92718	0.40403
23	0.39073	0.92050	0.42447
24	0.40674	0.91355	0.44523
25	**0.42262**	**0.90631**	**0.46631**
26	0.43837	0.89879	0.48773
27	0.45399	0.89101	0.50953
28	0.46947	0.88295	0.53171
29	0.48481	0.87462	0.55431
30	**0.50000**	**0.86603**	**0.57735**

TRIGONOMETRIC TABLES (30-60)

ANGLE	SIN	COS	TAN
31	0.51504	0.85717	0.60086
32	0.52992	0.84805	0.62487
33	0.54464	0.83867	0.64941
34	0.55919	0.82904	0.67451
35	**0.57358**	**0.81915**	**0.70021**
36	0.58779	0.80902	0.72654
37	0.60182	0.79864	0.75355
38	0.61566	0.78801	0.78129
39	0.62932	0.77715	0.80978
40	**0.64279**	**0.76604**	**0.83910**
41	0.65606	0.75471	0.86929
42	0.66913	0.74314	0.90040
43	0.68200	0.73135	0.93252
44	0.69466	0.71934	0.96569
45	**0.70711**	**0.70711**	**1.00000**
46	0.71934	0.69466	1.03553
47	0.73135	0.68200	1.07237
48	0.74314	0.66913	1.11061
49	0.75471	0.65606	1.15037
50	**0.76604**	**0.64279**	**1.19175**
51	0.77715	0.62932	1.23490
52	0.78801	0.61566	1.27994
53	0.79864	0.60181	1.32704
54	0.80902	0.58778	1.37638
55	**0.81915**	**0.57358**	**1.42815**
56	0.82904	0.55919	1.48256
57	0.83867	0.54464	1.53986
58	0.84805	0.52992	1.60033
59	0.85717	0.51504	1.66428
60	**0.86603**	**0.50000**	**1.73205**

TRIGONOMETRIC TABLES (60-90)

ANGLE	SIN	COS	TAN
61	0.87462	0.48481	1.80405
62	0.88295	0.46947	1.88073
63	0.89101	0.45399	1.96261
64	0.89879	0.43837	2.05030
65	**0.90631**	**0.42262**	**2.14451**
66	0.91355	0.40674	2.24604
67	0.92050	0.39073	2.35585
68	0.92718	0.37461	2.47509
69	0.93358	0.35837	2.60509
70	**0.93969**	**0.34202**	**2.74748**
71	0.94552	0.32557	2.90421
72	0.95106	0.30902	3.07768
73	0.95630	0.29237	3.27085
74	0.96126	0.27564	3.48741
75	**0.96593**	**0.25882**	**3.73205**
76	0.97030	0.24192	4.01078
77	0.97437	0.22495	4.33148
78	0.97815	0.20791	4.70463
79	0.98163	0.19081	5.14455
80	**0.98481**	**0.17365**	**5.67128**
81	0.98769	0.15643	6.31375
82	0.99027	0.13917	7.11537
83	0.99255	0.12187	8.14435
84	0.99452	0.10453	9.51436
85	**0.99619**	**0.08715**	**11.43005**
86	0.99756	0.06975	14.30067
87	0.99863	0.05233	19.08114
88	0.99939	0.03490	28.63625
89	0.99985	0.01745	57.28996
90	**1.00000**	**0.00000**	∞

Appendix C

Conversion Factors

ENGLISH TO METRIC CONVERSION FACTORS

Length

1 in.	0254 m
1 in.	2.54 cm.
1 in.	25.4 mm
1 in.	.083333 ft.
1 mil	.001 in.
1 mil	.0254 mm

Mass

1 lb.	.4536 kg
1 lb.	16 oz.
1 lb.	453.6 gm.
1 oz.	28.35 gm.

Velocity

1 rpm	.1047 rad./sec
1 ft/sec	.682 miles/hr.
1 cm/sec	1.97 ft/min.

Acceleration

1 ft/min/min	0084674 cm/sec/sec.
1 ft/sec/sec	.032808 cm/sec/sec

FRACTIONS TO DECIMALS

1/16	0.0625
1/8	0.125
3/16	0.1875
1/4	0.25
5/16	0.3125
3/8	0.375
7/16	0.4375
1/2	0.5
9/16	0.5625
5/8	0.625
11/16	0.6875
3/4	0.75
13/16	0.8125
7/8	0.875
15/16	0.9375

Appendix D

Weight Loss Due to Drilling

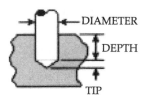

Figure D-1. Removing
Weight by Drilling

VOLUME OF MATERIAL REMOVED

Table D-1

DIAMETER	TIP	PER 1/8TH INCH AFTER TIP
1/16	.0000165	.0003835
1/8	.0001317	.0015339
3/16	.0004444	.0034515
1/4	.0010534	.0061359
5/16	.0020574	.0095874
3/8	.0035553	.0138058
7/16	.0056456	.0187913
1/2	.0084273	.0245437
9/16	.0119990	.0310631
7/8	.0164596	.0383495
11/16	.0219077	.0464029
3/4	.0284421	.0552233
13/16	.0361616	.0648107
7/8	.0451650	.0751650
15/16	.0555551	.0862864
1	.0674183	.0981748
1-1/16	.0808658	.1108301
1-1/8	.0959921	.1242524
1-3/16	.1128961	.1384418
1-1/4	.1316764	.1533981

To get the total volume removed, add the tip volume to the depth drilled as shown in Figure (D-1). Use the following chart to determine the weight removed.

Table D-2

MATERIAL	OUNCES PER CU. IN.
ALUMINUM	1.52778
COPPER	5.14815
IRON	4.16667
LEAD	6.57407
STEEL	4.52778

Example D-1

A flywheel 24 inches in diameter requires 1-1/2 inch-ounces of correction weight at 137 degrees. It is decided to balance the flywheel by removing material rather than by adding a correction weight. The flywheel is made of nodular iron. What is the depth of the drilling if a 9/16-inch bit is to be used?

Step 1. First, since material is to be removed, it is removed opposite where the correction weight would be added, or 180+137 = 317 degrees. If a hole were to be drilled 10 inches from the center, the required weight to be removed would be 1.5/10 or .15 ounces.

Step 2. Iron weighs approximately 4.16667 ounces per cubic inch, from Table (D-2); therefore, a total of .15/4.16667 or .035999 cubic inches must be removed. From Table (D-1), the tip of the 9/16-in. bit removes .0119990 cubic inches. The additional depth required is thus .035999 − .011999 = .024 inches

Step 3. From Table (D-1) a 9/16-in. bit removes .0310631 cubic inches per 1/8 inch past the tip. Therefore, the number of 1/8-inch increments is .024/.0310631 = .7726 x .125 or .0966 inches past the tip.

Since this depth would be difficult to measure, a better solution may be to use a 1/2-inch bit. The 1/2-inch tip removes .0084273 cubic inches leaving .035999 - .0084273 or .02757 cubic inches. The 1/2-inch bit removes .0245437 per 1/8 inch or .125 inch of depth, and thus the required depth is (.02757/.0245437) x .125 or .14 inches.

Also note that multiple holes can be drilled at different radiuses and/or the removed weight can be broken into parts as with the correction weights.

Appendix E

Properties of Steel Shafts

FOR SOLID STEEL SHAFTS

Dia. in.	Moment of Inertia	Lbs. per Foot	Dia. in.	Moment of Inertia	Lbs. per Foot
0.25	0.0002	0.16	8.25	227.3975	178.19
0.5	0.0031	0.65	8.5	256.2392	189.15
0.75	0.0155	1.47	8.75	287.7412	200.44
1	0.0491	2.62	9	322.0623	212.06
1.25	0.1198	4.09	9.25	359.3659	224.00
1.5	0.2485	5.89	9.5	399.8198	236.27
1.75	0.4604	8.02	9.75	443.5968	248.87
2	0.7854	10.47	10	490.8739	261.80
2.25	1.2581	13.25	10.25	541.8329	275.05
2.5	1.9175	16.36	10.5	596.6602	288.63
2.75	2.8074	19.80	10.75	655.5469	302.54
3	3.9761	23.56	11	718.6884	316.78
3.25	5.4765	27.65	11.25	786.2850	331.34
3.5	7.3662	32.07	11.5	858.5414	346.23
3.75	9.7072	36.82	11.75	935.6671	361.45
4	12.5664	41.89	12	1017.8760	376.99
4.25	16.0150	47.29	12.25	1105.3867	392.86
4.5	20.1289	53.01	12.5	1198.4225	409.06
4.75	24.9887	59.07	12.75	1297.2110	425.59
5	30.6796	65.45	13	1401.9848	442.44

(Continued)

FOR SOLID STEEL SHAFTS (*Cont'd*)

Dia. in.	Moment of Inertia	Lbs. per Foot	Dia. in.	Moment of Inertia	Lbs. per Foot
5.25	37.2913	72.16	13.25	1512.9808	459.62
5.5	44.9180	79.19	13.5	1630.4406	477.13
5.75	53.6588	86.56	13.75	1754.6104	494.96
6	63.6173	94.25	14	1885.7410	513.13
6.25	74.9014	102.27	14.25	2024.0878	531.62
6.5	87.6241	110.61	14.5	2169.9109	550.43
6.75	101.9025	119.28	14.75	2323.4749	569.58
7	117.8588	128.28	15	2485.0489	589.05
7.25	135.6194	137.61	15.25	2654.9068	608.85
7.5	155.3156	147.26	15.5	2833.3269	628.97
7.75	177.0829	157.24	15.75	3020.5924	649.43
8	201.0619	167.55	16	3216.9909	670.21

Alignment Forms

The forms in this section may be copied without permission. They will be found useful in the alignment process as well as in maintaining maintenance records.

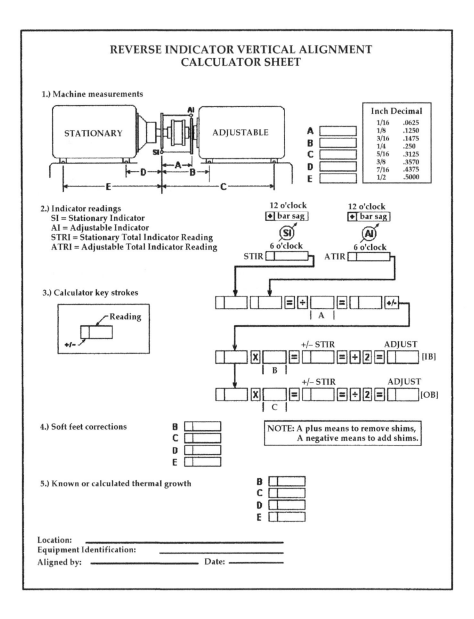

REVERSE INDICATOR VERTICAL ALIGNMENT CALCULATOR SHEET

1.) Machine measurements

Inch Decimal	
1/16	.0625
1/8	.1250
3/16	.1475
1/4	.250
5/16	.3125
3/8	.3570
7/16	.4375
1/2	.5000

STATIONARY   ADJUSTABLE

A B C D E

2.) Indicator readings
SI = Stationary Indicator
AI = Adjustable Indicator
STRI = Stationary Total Indicator Reading
ATRI = Adjustable Total Indicator Reading

12 o'clock   12 o'clock
bar sag   bar sag
SI   AI
6 o'clock   6 o'clock
STIR   ATIR

3.) Calculator key strokes

Reading
+/−

| A |

+/− STIR   ADJUST
X   =   =+2=   [IB]
| B |

+/− STIR   ADJUST
X   =   =÷2=   [OB]
| C |

4.) Soft feet corrections   B
C
D
E

NOTE: A plus means to remove shims,
A negative means to add shims.

5.) Known or calculated thermal growth   B
C
D
E

Location: _____
Equipment Identification: _____
Aligned by: _____ Date: _____

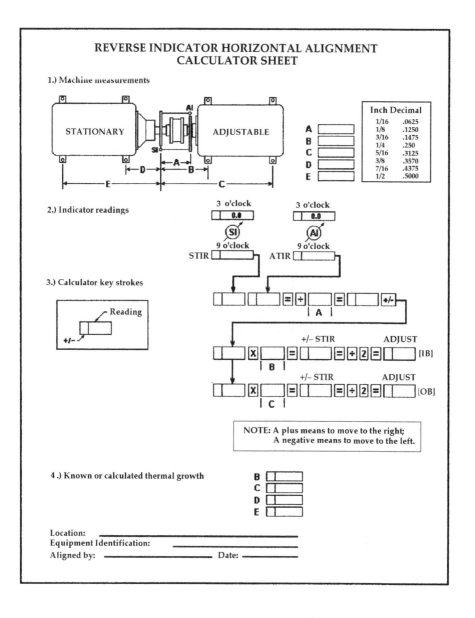

REVERSE INDICATOR HORIZONTAL ALIGNMENT
CALCULATOR SHEET

1.) Machine measurements

STATIONARY

ADJUSTABLE

Inch Decimal	
1/16	.0625
1/8	.1250
3/16	.1475
1/4	.250
5/16	.3125
3/8	.3570
7/16	.4375
1/2	.5000

2.) Indicator readings

3 o'clock
0.0

3 o'clock
0.0

9 o'clock
STIR

9 o'clock
ATIR

3.) Calculator key strokes

Reading
+/–

= ÷   =   +/–
| A |

+/– STIR   ADJUST
X   =   = ÷ 2 =   [IB]
| B |

+/– STIR   ADJUST
X   =   = ÷ 2 =   [OB]
| C |

NOTE: A plus means to move to the right;
A negative means to move to the left.

4.) Known or calculated thermal growth
B
C
D
E

Location:
Equipment Identification:
Aligned by: Date:

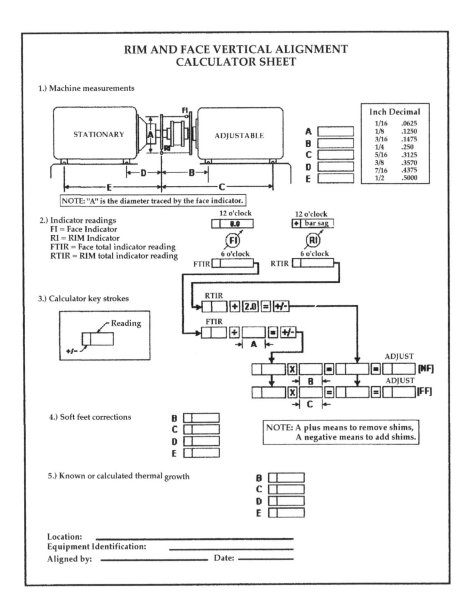

RIM AND FACE VERTICAL ALIGNMENT CALCULATOR SHEET

1.) Machine measurements

Inch Decimal	
1/16	.0625
1/8	.1250
3/16	.1475
1/4	.250
5/16	.3125
3/8	.3570
7/16	.4375
1/2	.5000

STATIONARY   ADJUSTABLE

NOTE: "A" is the diameter traced by the face indicator.

2.) Indicator readings
FI = Face Indicator
RI = RIM Indicator
FTIR = Face total indicator reading
RTIR = RIM total indicator reading

3.) Calculator key strokes

Reading

ADJUST [NF]
ADJUST [FF]

4.) Soft feet corrections

NOTE: A plus means to remove shims, A negative means to add shims.

5.) Known or calculated thermal growth

Location:
Equipment Identification:
Aligned by: Date:

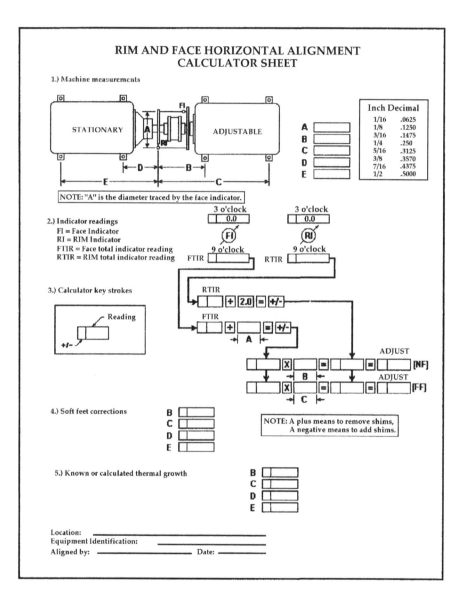

RIM AND FACE HORIZONTAL ALIGNMENT
CALCULATOR SHEET

1.) Machine measurements

STATIONARY

ADJUSTABLE

Inch Decimal	
1/16	.0625
1/8	.1250
3/16	.1475
1/4	.250
5/16	.3125
3/8	.3570
7/16	.4375
1/2	.5000

NOTE: "A" is the diameter traced by the face indicator.

2.) Indicator readings
FI = Face Indicator
RI = RIM Indicator
FTIR = Face total indicator reading
RTIR = RIM total indicator reading

3.) Calculator key strokes

Reading

ADJUST [NF]
ADJUST [FF]

4.) Soft feet corrections
B
C
D
E

NOTE: A plus means to remove shims,
A negative means to add shims.

5.) Known or calculated thermal growth
B
C
D
E

Location:
Equipment Identification:
Aligned by: Date:

SOFT FEET CORRERCTION SHEET

1.) Machine measurements

STATIONARY   ADJUSTABLE

a
b
c
d
Minimum (a-d)   S1

e
f
g
h
Minimum (e-h)   A1

2.) Soft feet corrections

> Subtract S1 from readings a-d, then enter in A-D below.
> Subtract A1 from readings e-h, then enter E-H below.

A   E
B   F
C   G
D   H

Location: _____
Equipment Identification: _____
Aligned by: _____ Date: _____

Procedures

This section contains sample procedures for cooling tower fans. The various procedures are easily modified to fit other types of equipment, and should be used as guidelines in the development of other procedures.

In addition, the following flow charts are provided with the same intent. These procedures and flow charts may be copied in whole or in part to develop specific procedures for other equipment.

Prior to performing any maintenance or procedures, safety rules and regulations should be reviewed. A generalized safety procedure is also provided. The following illustration shows the logical flow of maintenance for a cooling tower fan.

The following flow charts are presented as amplifications of the segments shown in Figure (P-1).

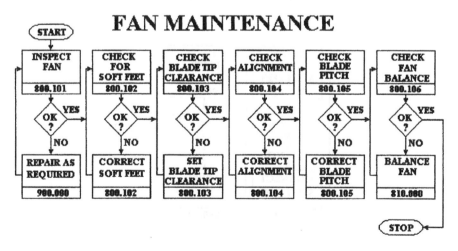

Figure P-1. Fan Maintenance Flow Chart

INSPECTIONS

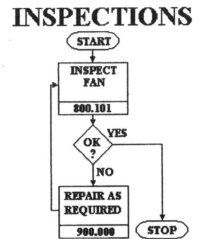

Figure P-2. Inspection Flow Chart

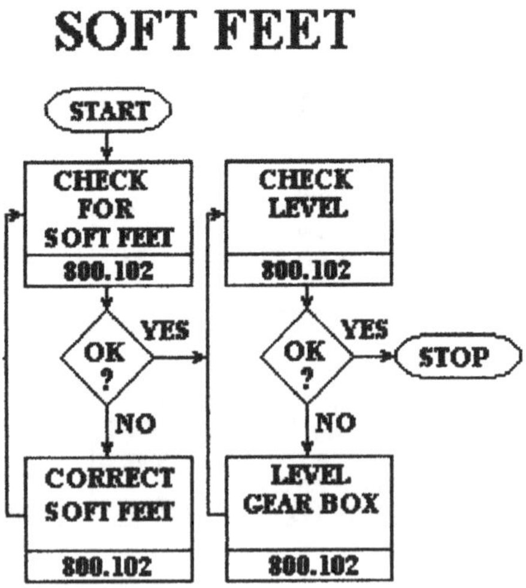

SOFT FEET

Figure P-3. Flow Chart for Soft Feet Corrections

BLADE CLEARANCE

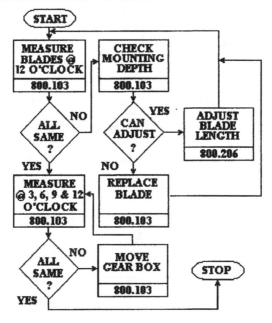

Figure P-4. Flow Chart for Blade Clearance

ALIGNMENT

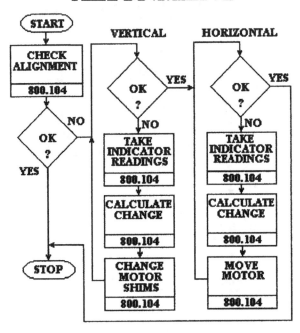

Figure P-5. Flow Chart for Fan Alignment

ADJUST PITCH

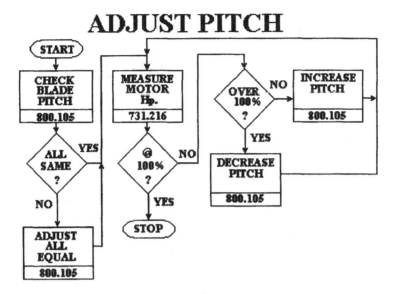

Figure P-6. Flow Chart for Adjusting Blade Pitch

BALANCE

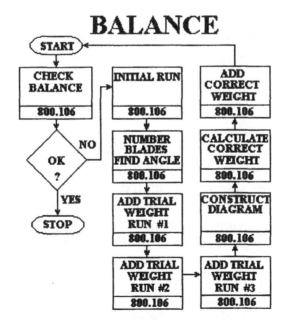

Figure P-7. Flow Chart for Balancing Slow Speed Fans

Procedure 800.000

Safety Notes

PURPOSE

This procedure addresses general safety practices and cautions which must be reviewed and understood to assure safety when performing maintenance or inspections of cooling towers or fin fan coolers. It is not the intended purpose of this procedure to be totally encompassing in its scope, but rather to supplement established safety practices, policies and procedures, both general and specific to the area or facility.

RECOMMENDED INTERVAL

These guidelines must be adhered to when performing maintenance or inspections of cooling towers and fin fan coolers.

GENERAL

Prior to performing maintenance or inspections of any equipment, an understanding of safety procedures is required. Become familiar with any special rules and/or regulations that may apply to the specific area of the plant or facility that you will be working in. Thoroughly review all emergency procedures. Often maintenance personnel perform work in many different areas or locations and are not sufficiently familiar with emergency operational procedures to be of assistance.

NOTE: It is advisable that a brief safety meeting be held each morning to assure all personnel working in the area are aware of the scheduled activities. This should include designating a location to meet in the event of an emergency to ascertain an accurate head count. In addition, discuss what assistance each individual will provide in handling the emergency.

PROCEDURE

1. Place safety first!
2. Read and understand all procedures to be performed, and adhere to all cautions, warnings, and notes on safety.
3. Review lock out and tag out procedures.
4. Assure operating personnel have all been properly notified of your planned activities.
5. Safety glasses, hard hat, and other protective equipment stipulated for the work location, must be worn at all times.
6. Observe and adhere to all safety warning signs, such as NO SMOKING.
7. Inspect stairways, ladders and railings for wood rot, secure fasteners, and general structural integrity.

CAUTION: Large cooling towers must have two means of exiting. In general, most towers have a stairway and either a ladder or other escape system which should be located at the opposite end of the tower. Most towers are fabricated of flammable materials and often the fluid(s) being cooled are flammable. A second escape route is imperative in the event of a fire on or near the primary means of access to the tower.

8. Assure a fire extinguisher is available on the cooling tower deck, and that it has been properly tested and inspected.
9. Sniff (using the proper sniffer or other instrument) for gas or other process fluid leaks, which could be either toxic and/or flammable.
10. When the process fluid is toxic, assure breathing apparatuses are available and in proper working order.

CAUTION: When working alone, assure someone is advised of your location, and an expected time you will check in.

11. Assure all guards and shields in place and properly secured at all times except when making adjustments.
12. Be aware of other operational and/or maintenance activities that could interfere with your safety or the safety of others.
13. Watch for inclement weather, which could produce high winds and/or lightning in the area.

WARNING: It may be necessary to tie down certain equipment and parts on towers where wind currents could blow them off and cause injury to persons or equipment below.

14. Do not enter the tower without the use of a safety harness in the event of a slip.

CAUTION: It is advisable to have a second person present when entering the fan area or cooling tower is necessary.

15. Assure hoists or other lifting devices used to remove equipment or lift equipment onto the tower are properly rated, inspected to assure the integrity of cables, properly secured to the tower, and are in good working order.
16. Assure the integrity of the decking in and around your work area. Additional supports may be required when removing heavy equipment such as motors and gearboxes.
17. Use proper rigging procedures to secure all loads.
18. Establish clear and precise hand signals, to communicate to persons on the ground.
19. Clean any oil spills immediately to prevent slip and fall and a fire hazard.
20. Keep all hand tools not in use in a tool box to avoid trip hazards.
21. Test all electrical wires and motor frames to assure there is no stray voltage.

22. Use only recently calibrated test equipment.
23. Assure all moving parts are secure, including any weight added to fan blades, prior to operating the equipment.

CAUTION: Loose objects can become high-speed projectiles and cause serious injury!

24 Inspect all extension cords for insulation, plugs, proper grounding and proper wire gauge for the intended job.
25. Pay attention to the routing of electrical cables.

CAUTION: It may be advisable to cover cables in traffic areas to prevent trips.

26. Keep path of cables as short as possible.
27. Unused cables and extension cords should be coiled and stored out of the way.
28. Double check to assure no tools or rags are left in or around the fan stack or shroud prior to untagging and unlocking the power to the motor.
29. Thoroughly inspect your work area to assure it is left in a safe manner.
30. Report any equipment or structural members that are unsafe, to the proper authority.
31. Keep proper written notes. Do not rely on memory!

Fan Inspections

PURPOSE

This procedure addresses the basic inspection procedures for fans and their associated structural members, prior to performing major maintenance.

RECOMMENDED INTERVAL

Semi-annually or prior to performing major maintenance.

GENERAL

Prior to performing major maintenance of cooling fans, general inspections are required. The mechanical condition of the supporting members and all fastening devices must be determined. In addition, the inspections must note any internal and external corrosion and the buildup of scale and other debris. The proper lubrication and general condition of seals and bearings must also be noted.

I. PROCEDURE 800.101.1

Fan Deck Inspections
1. Inspect the fan deck for decay, missing and/or broken members.
2. Inspect for gaps between decking which could cause short circuiting of the air circulation.
3. Towers with metal decking must be inspected for holes in the decking due to corrosion. The condition of the corrosion protective coating must be noted.
4. Inspect the fan deck supports for decay and loose or broken members.

5. All fasteners must be inspected to assure they are tight and free of excessive corrosion.
6. Record all findings.

II. PROCEDURE 800.101.2

Fan Stack Inspections
1. Fan stacks must be inspected to assure they are securely attached to the fan deck supports. Check for looseness of fasteners and assure they are not excessively corroded.
2. Check for any missing and/or broken members of the stack structure, and note any deterioration of any member.
3. Observe any wood rot and corrosion of metal retaining rings on wooden stacks.
4. Inspect tie down and support straps on fiberglass stacks.
5. Inspect for proper blade tip clearance using **Procedure 800.102**.
6. If so equipped, inspect fan guards for loose fasteners and excessive corrosion. Check to assure no loose or broken members could fall into the operating fan.

NOTE: OSHA requires all fans with stack heights of less than five feet tall be equipped with a fan guard over the entire stack.

7. Record all findings.

III. PROCEDURE 800.101.3

Mechanical Equipment Support Inspections
1. Insect the equipment support members for corrosion, rot, loose or missing fasteners, straps and/or hangers.

CAUTION: Prior to entering the fan stack area, assure the motor is locked out and tagged out. In addition, tie down the fan blades to assure accidental rotation cannot occur due to natural drafts within the tower.

2. Inspect any wooden members in contact with equipment supports to ensure structural soundness. Check wood for iron rot in areas where wood contacts metal structures.
3. If vibration dampers are used, such as springs or rubber pads, check for corrosion and/or loss of vibration absorption properties.
4. Assure all hold-down fasteners are properly torqued.
5. Record all findings.

IV. PROCEDURE 800.101.4

Fan Inspections
1. Inspect the fan hub for signs of corrosion and assure all fasteners are in good condition and properly torqued.
2. Inspect blade clamping devices for signs of corrosion and assure they are properly torqued.
3. Inspect each fan blade for corrosion and erosion. Blades must be inspected for any build-up of solid materials which can alter both the air foil characteristics and the balance of the fan.

NOTE: Hollow blades should have a drain hole drilled near their tip to assure no accumulation of moisture that can alter the fan balance. Assure tall drain holes are open.

4. Clean the blades of all foreign material.
5. Inspect the blades for proper pitch angle, using **Procedure 800.105.6**.
6. Record all findings.

V. PROCEDURE 800.101.5

Gearbox Inspections
1. Check the gearbox for proper oil level and fill as required, using the proper oil.

2. Inspect the gearbox oil for signs of moisture.
3. Collect a small sample of the oil on a white paper towel and inspect under a strong light for metallic wear particles.

NOTE: Gearbox oil should be changed annually or after extended periods of down time.

4. Rotate the fan back and forth to determine the amount of play in the gear teeth.
5. Check the input shaft bearing by attempting to move the shaft in a radial direction.
6. Inspect the fan shaft bearing for excessive end play by attempting to lift the fan blade while noting any movement of the output shaft.
7. Record all findings and action taken.

VI. PROCEDURE 800.101.6

Driveshaft & Coupling Inspections
1. Inspect the drive shaft for any signs of corrosion. The condition of the internal surface of the drive shaft can be determined by lightly tapping the drive shaft with a small hammer, along its entire length.

NOTE: Areas that produce a dead sound indicate possible internal corrosion.

2. Inspect all keys to assure they are not excessively corroded and that they are of the proper dimensions to assure proper balance. See Couplings, and refer to **Procedure 800.106**.
3. All fastening devices must be inspected for excessive corrosion and checked to determine they are properly torqued. Check to assure all fasteners are of the same grade and the same length. Assure all washers are in place and of the same dimensions, if so equipped.

4. Inspect couplings for excessive corrosion, wear and proper lubrication.
5. Inspect shim pack type couplings for cracking, broken or missing members.
6. Inspect U-joint couplings for proper lubrication.

NOTE: These types of couplings must operate with the proper amount of misalignment to assure proper lubrication of the U-joint bearing needles. Excessive misalignment causes excessive velocity changes and loss in efficiency. See **Procedure 800.101.6**.

7. Inspect motor coupling guard to assure it is properly fastened down. Inspect all fasteners for excessive corrosion.

NOTE: The condition of this guard is important in the prevention of serious injury and/or excessive damage to equipment in the event of a coupling or drive shaft failure!

7. Record all findings and any action taken.

VII. PROCEDURE 800.101.7

Motor Inspections
1. Inspect the motor for signs of over heating and hot spots.
2. Assure the bearings are properly lubricated.

NOTE: Over lubrication of motor bearings may lead to insulation deterioration.

3. Inspect the motor air inlet passage for excessive build-up of dirt and other solid debris, and clean if required.
4. Inspect the motor fan to assure all blades are clean and not broken or damaged. Clean or replace as required.
5. If the motor is equipped with a moisture drain, assure it is functioning properly.
6. Examine all fasteners for corrosion, and assure they are properly torqued.
7. Record all findings and action taken.

Soft Feet Corrections

PURPOSE

This procedure addresses the requirements for determining uneven feet on both drivers and fans, and outlines the required steps to correct them. This will assure proper internal equipment alignment, and ease the effort required for coupling alignment. Coupling, bearing, seal and gear failures will be reduced.

RECOMMENDED INTERVAL

After major component replacement or when equipment is re-aligned.

GENERAL

Prior to attempting to properly align equipment, soft feet must be measured and corrected. Failure to compensate for soft feet could result in internal misalignment and lead to premature failures. Gearboxes should be leveled to assure the lack of aerodynamic forces and to assist in the alignment process. Aerodynamic forces can be generated from uneven blade passage over internal supports within the fan stack.

PROCEDURE

1. Lock out and tag out the fan motor power.
2. Lock Loosen all hold-down bolts. Remove, measure and record the amount of shim material under each foot.
3. Lock Clean all foreign material from under the gear box feet, such as paint, grease, oil and dirt.
4. Lock Torque all hold-down bolts to their proper tightness.
5. Lock Using a felt tip marker or similar device, label each

foot. Facing the gearbox input shaft, note "left" and "right."

6. Lock Measure and record the machine soft feet by one of the following methods.

I. Dial Indicator Method
A. Mount a dial indicator to the base plate so the stem will read the vertical travel of the foot as the hold-down bolt is loosened.
B. Loosen one hold-down bolt at a time noting and re-cording the indicator reading. Re-torque the hold-down bolt to its proper value.
C. Repeat the steps above for the remaining feet.

II. Feeler Gauge Method
A. Loosen a hold-down bolt, and using a feeler gauge, determine the clearance between the foot and the support member. Record this information.
B. Re-torque the hold-down bolt to its proper value.
C. Repeat the steps above for the remaining feet.

7. Subtract the smallest reading from the remaining readings for each foot. The results are the proper corrections for each foot.
8. Using proper shim material, add the proper amount of shims under each foot.
9. Re-torque the hold down bolts and record the corrections for future reference
10. Use a machinists or carpenters level to assure the gearbox is properly leveled.

    A. If the gear box is not level then perform the following:
    i. Use shims or a feeler gauge under the level to deter-mine the amount of correction over the length of the level.
    ii. Measure the distance between the hold-down bolts.

iii.   Determine the correct amount of shims to add under the low foot, using the following formula:

$$C = R * H/L \qquad\qquad\qquad\text{(P-1)}$$

Where:
   C = the amount of correction.
   R = the shims or feeler gauge reading under the
level.
   H = the distance between the hold-down bolts.
   L = the length of the level.

NOTE: The gearbox must be leveled in both planes. After correcting the level from front to rear, correct the level from side to side.

iv.   Record the amount of shims added under each foot.

Procedure 800.103

Blade Tip Clearance

PURPOSE

This procedure addresses the requirements for determining the tip clearance of fan blades, and the steps for moving the gearbox to assure proper tip clearance. This assures the fan will operate without excessive aerodynamic forces that can cause excessive vibration.

RECOMMENDED INTERVAL

Annually or after gear box or fan blade replacement.

GENERAL

Prior to measuring the blade tip clearance, complete an inspection of support members and fastening devices. In addition, the gearbox must be leveled and any soft feet corrected. See **Procedures 800.102 and 800.102**.

Fan hubs and fan blades also must be cleaned of any scale or other buildup. The fan stack must be in a good state of repair and rigid. If so equipped, fan guards may need to be removed prior to performing this procedure, and must be replaced when completed.

CAUTION: Fan blades should be secured at all times except when rotating them to make measurements. Natural drafts through the tower can cause blades to rotate that could lead to injury.

PROCEDURE

1. Lock and tag out motor to fan, and secure fan blades.
2. Randomly select a blade and using a felt tip marker label it number one. Continue in a counter clockwise direction and number the remaining blades
3. Using the drive shaft as a reference point, place a mark on the fan stack near where the blades pass and label it 12 o'clock. Locate and label the 3, 6 and 9 o'clock positions on the fan stack, where the blade tip passes.
4. Using a steel ruler or a tape measure, measure and record the distance between the closest point on each fan blade and the 12 o'clock mark on the fan stack.
5. Rotate the fan and measure the clearance of the number one blade at the 3, 6, and 9 o'clock positions, as in step 4.
6. From the readings taken in step 4, determine if any blade is either too short or too long, and inspect the mounting at the hub to determine if it is properly seated.
   a. If the blade is not seated properly, correct the seating depth by loosening the retaining bolts and moving the blade either in or out to achieve the proper blade length. Re-torque the retaining bolts and repeat step 4.
   b. If the blade cannot be corrected as in step 6A, it may need to be replaced.

7. Subtract the 6 o'clock reading from the 12 o'clock reading, and divide this result by two. This determines the amount of required movement in the plane parallel to the drive shaft.

NOTE: A positive number indicates the gearbox must be moved toward the 12 o'clock mark, a negative number indicates it must be moved toward 6 o'clock.

8. Subtract the 9 o'clock reading from the 3 o'clock reading, and divide this result by two. This determines the amount

of required movement in the plane perpendicular to the drive shaft.

NOTE: A positive number indicates the gearbox must be moved toward the 9 o'clock mark; a negative number indicates it must be moved toward the 3 o'clock.

9. Attach dial indicators to support members so that they can read the movement of the gearbox. Zero the indicators, loosen the hold-down bolts and bump the gearbox in the proper direction, noting the amount of movement.

NOTE: An alternate method is to locate the number-one blade facing the direction in which the gearbox needs to be moved, and measure the tip clearance after each bump.

9. Torque all hold-down bolts and re-measure the tip clearances. Repeat steps 1 through 9 as required.
10. Use **Procedure 800.104** to re-align the drive shaft couplings.

Procedure 800.104

Fan Alignment

PURPOSE

This procedure addresses the requirements for proper alignment of the motor and fan gearbox of cooling tower fans, or other large slow-speed fans using the rim and face alignment method. This assures the couplings are properly aligned and that vibration due to misalignment will be minimized. In addition, this assures prolonged life of couplings, seals, bearings and gears.

RECOMMENDED INTERVAL

After major component replacement or when excessive vibration is noted.

GENERAL

Prior to attempting to properly align cooling tower fans, assure all inspections have been properly made. Refer to **Procedure 800.100**. In addition, assure fan is properly leveled and entered within the fan shroud. Also, assure soft feet have been corrected. Refer to **Procedures 800.102 and 800.103**.

CAUTION: This procedure must not be used on fans with U-joint couplings. Refer to **Procedure 800.104.1** for U-joint couplings.

PROCEDURE

1. Lock out and tag out the motor to the fan.

CAUTION: Fan blades should be secured at all times, except when rotating the indicator fixtures. Natural draft through the tower can cause blades to rotate that could lead to injury.

2. Setup indicator fixture on the gearbox coupling to determine fixture measurements. Mark the exact location where the fixture mounts to the gearbox, using a felt-tip marker or similar device.
3. Remove fixture, and set up on a ridged pipe using the same dimensions and fixture components to be used during the alignment process.
4. Place the rim indicator at the 6 o'clock position and zero the indicator.
5. Rotate the fixture to place the rim indicator in the 12 o'clock position. Read and record the bar sag.
6. Assure all motor and gear box feet have been corrected for soft feet. Refer to **Procedure 800.102**.
7. Assure gear box is level and centered in the fan shroud. Refer to **Procedure 800.103**.
8. Place the alignment fixture on the gearbox coupling so that the indicators read on the drive shaft side of the coupling. Use the marks to assure the fixture is configured properly for bar sag compensation.
9. Measure and record the A, B1 and C1 distances.
10. Set the face indicator to zero, and the rim indicator to the [+] plus bar sag reading at 12 o'clock. Rotate the drive shaft in the direction of normal operation, one complete revolution to assure the indicator's return.
11. Rotate the drive shaft in the direction of normal operation to place the rim and face indicators in the 6 o'clock position, then take and record the indicator readings.

CAUTION: Always observe the indicators during the rotation of the fixture to assure the readings do not pass back through zero. This can cause severe errors if a small reading was interpreted as a large reading in the opposite direction.

12. Rotate the drive shaft to the 12 o'clock position to assure the indicators return to their initial readings. If they do not, tighten the fixture elements and repeat steps 10 through 12.

13. Remove the alignment fixture and install it on the motor coupling so that the indicators read on the motor side of the coupling. Assure the fixture dimensions remain the same. Mark the location of the fixture on the drive shaft for future installations.

14. Measure the new B2 and C2 distances and record them.

15. Repeat steps 10 through 12 for the new fixture location. Read and record the new indicator readings.

16. Determine the required shim changes for the motor feet, using the graphical method, calculator method or the computer program.

17. Verify all shims by using a micrometer. Remove any ridges or burrs from shim material.

18. Repeat steps 4 through 17 as required to correct the vertical alignment

19. Once the vertical alignment is corrected, mark the LEFT side of the machine, using a felt-tip marker.

NOTE: Although either side of the machine can be designated as the left side without causing errors, it is easier if the motor is marked as viewed from the gearbox end.

20. Locate the fixture on the gearbox coupling, using the marks on the shaft to assure the fixture is placed in its original position.

21. Rotate the drive shaft in the direction of normal operation to align the indicators on the left or 9 o'clock position. Set the indicators to zero.

NOTE: Equipment with gears can cause shifts in readings if not rotated in the direction of normal operation, due to gear backlash. Never go past the reading location and rotate back to it. If rotated past the desired position, continue a complete rotation in the proper direction to reach the reading location.

22. Rotate the drive shaft one full revolution in the direction of normal operation to assure the indicators repeat their original readings.
23. Rotate the drive shaft to the 3 o'clock position and read and record the indicator readings.
24. Return the fixtures to the o'clock position to assure they return to zero.
25. Move the indicator fixture to the marked location on the motor coupling.
26. Repeat steps 21 through 24, and record the indicator readings.
27. Determine the required motor movements. (See step 16)
28. Repeat steps 20 through 26 as required.
29. Torque all hold-down bolts and assure all equipment is clear and ready for operation.
30. Remove lock and tag and place unit in operation.
31. Measure and record the final vibration levels.

Procedure 800.104..1

Alignment of
U-joint Couplings

PURPOSE

This procedure addresses the requirements for proper alignment of the motor and fan gearbox of cooling tower fans, or other
large slow-speed fans. This assures the couplings are properly
aligned and that vibration due to misalignment will be minimized. In addition, this assures prolonged life of couplings, seals,
bearings and gears.

RECOMMENDED INTERVAL

After major component replacement or when excessive vibration is noted.

GENERAL

Prior to attempting to properly align cooling tower fans, assure all inspections have been properly made. Refer to **Procedure
800.100**. In addition, assure fan is properly leveled and entered
within the fan shroud. Also, assure soft feet have been corrected.
Refer to **Procedures 800.102 and 800.103**.

NOTE: To achieve proper lubrication of U-joint bearing needles, a
minimum of 1 to 3 degrees of angularity is required. The maximum angularity allowed is 7 to 8 degrees, to avoid loss of effi-

ciency and large velocity changes, which occur at two times drive shaft RPM. This procedure aligns the U-joint coupling to approximately 5 degrees.

Often lack of lubrication in U-joints is blamed on corrosion and not the root cause, proper alignment to assure proper lubrication of the needles.

PROCEDURE

1. Lock out and tag out motor, and secure the fan blades.
2. Assure the gearbox and motor are leveled.
3. Place a large level across the gearbox coupling and mark off two locations on the drive shaft one foot apart.
4. Place one end of the level on one mark and hold it in a level position.
5. Measure and record the distance from the second mark to the bottom of the level. This distance must be 1 inch +/- 1/4 inch.
6. If the reading is not within the specified range, the motor must be moved.
7. Measure the distance between the centers of the two U-joint couplings.
8. Determine the amount of motor movement using the following formula:

$$S = (L/12) * (1 - D) \hspace{3cm} \text{(P-2)}$$

Where:
   S   = The amount of motor movement required in inches.
   L   = Length of the drive shaft in inches.
   D   = The distance from the level to the drive shaft.

9. Determine the proper direction, and move the motor.
10. Follow steps 20 through 30 of **Procedure 800.104** for proper horizontal alignment.

Procedure 800.105

Blade Pitch Adjustment

PURPOSE

This procedure addresses the requirements for determining the consumed horsepower of fan motors and the regulation of the load by adjustment of the fan blade pitch. This optimizes motor efficiency and power consumption.

RECOMMENDED INTERVAL

After blade adjustment or major component replacement.

GENERAL

Prior to measuring the consumed horsepower, the pitch angle of each blade must be measured, and set to the same pitch. It is best to have a bubble protractor to accomplish setting the pitch angle, but a carpenter or machinist level and a standard protractor can be used. A clamp-on ammeter and a voltmeter are required to measure the motor horsepower.

PROCEDURE

1. Lock and tag out motor to the fan, and secure the fan blades.
2. Using a black felt-tip marker or similar device, umber the blades in a counter clockwise direction (1, 2, 3, etc.), starting with any blade.
3. Place the level on the trailing edge of the blade marked number one, perpendicular to the axis of the fan blade and adjust it to level. Place the protractor on the fan blade from

the leading edge to the trailing edge, and measure and record the angle it forms.

NOTE: When using a small protractor, it may be necessary to use a flat strip of steel to entirely span the fan blade.

4. Repeat steps 2 through 3 for the remaining blades.
5. Find the average angle, and adjust all blades to that value.
6. Loosen the retaining bolts and rotate the blade while observing the angle.

CAUTION: Assure the blade seating depth is not altered while rotating blades. Mark the seating depth with a felt marker prior to loosing the bolts.

7. Re-torque all clamping bolts.
8. Assure all tools and personnel are clear of the fan, then remove the lock and tag and start the unit.
9. Record the following motor nameplate information:
   A. Motor Efficiency.
   B. Power Factor.
   C. Voltage.
   D. Rated Amperes.
   E. Number of Phases.

10. Start the unit and measure and record the actual voltage and amperes delivered to the motor.
11. Calculate the motor horsepower with one of the following formulas.
   A. SINGLE PHASE AC MOTORS:
   $$BHp = (I * E * EFF * PF)/746 \qquad (P-3)$$

   B. THREE PHASE AC MOTORS:
   $$BHp = (1.73 * I * E * EFF * PF)/746 \qquad (P-4)$$
   Where:
   I = Measured Amperes.
   E = Measured Voltage.
   EFF = Nameplate Efficiency.
   PF = Nameplate Power Factor.

12. Compare the measured horsepower to the nameplate horsepower to determine the percent load, by using the following formula:

PERCENT LOAD = (BHp/MHP)/100   (P-5)

Where:

    BHp = Nameplate Rated Horsepower.

    MHp = Measured Horsepower.

13. If the percent load is greater than 100% the blade pitch must be decreased; if it is less than 95%, it should be increased.

NOTE: If the supplied voltage is lower than the nameplate voltage, the rated amperage should also be reduced. Use the following formula:

$$DRA = (MV/NPV) * NPA \qquad (P-6)$$

  Where:

    DRA = De-rated Amperes.

    MV = Measured Voltage.

    NPV = Nameplate Voltage.

    NPA = Nameplate Amperage.

CAUTION: In general, small pitch changes can cause large changes in the required horsepower. Exercise care in adjusting the blade pitch. A general guide is that 1 degree of pitch changes the horsepower by 8%. Since this is only a rough estimate, manufacturers' fan curves should be consulted for more exact figures.

14. If any pitch changes are made, re-measure the percent of motor load.

CAUTION: Motors should not be operated over 100% load. To do so could over heat the motor windings and cause a failure. This is true of motors that have 110% or 115% service factors, due to potential fluctuations in loads.

Procedure 800.106

Fan Balancing

PURPOSE

This procedure addresses the requirements for correcting conditions of unbalance in cooling fans. This assures the unit will operate with minimum amounts of vibration, which can cause premature failures of gears, seals, bearings and couplings.

RECOMMENDED INTERVAL

After major component replacement or when excessive vibration is noted.

GENERAL

Prior to attempting to balance a fan, the pitch angle of each blade must be set; the gear box and motor must be corrected for soft feet; the unit must be properly aligned; and all mechanical components must be in a good state of repair. Refer to **Procedure 800.101** for inspections that must be made prior to balancing the unit. This procedure describes a method for field balancing of a fan, and requires no measurement of phase angle. Only amplitude measurements are required.

PROCEDURE

1. Lock out and tag out the fan motor.
2. Securely mount a vibration pickup to the fan gearbox in the radial plane.
3. Determine the fan RPM from the motor nameplate and the gear ratio of the fan gear box.

4. Set the vibration instrument to the filter-in position and adjust the frequency to the operating RPM of the fan.

5. Assure rotating parts of the equipment is clear of all tools, cables or other material, then remove the lock and tag.

6. Start the unit and record the amplitude of vibration present, by fine tuning the vibration instrument to maximize the reading at operating RPM.

NOTE: Since many large fans operate at relatively low speeds (200 - 400 RPM), most vibration pickups will be below their linear output range and may read amplitudes significantly below the actual vibration levels. This will not affect the balancing process, but does affect the balancing tolerance. The final balance should be corrected to assure the unit is within tolerance!

7. Stop the unit and lock out, tag out the motor and secure the fan blades.

8. Using a black felt-tip marker or similar device, number the blades in a counterclockwise direction (1, 2, 3, etc.), starting with any blade.

9. Attach a trial weight to the number one blade.

NOTE: Generally, a 2-ounce weight at the blade tip is sufficient for large slow speed fans, when attached to the blade tip. The proper size trial weight can be determined using the following formula:

$$Tw = .01 * RW/(.177 * (RPM/1000)^2 * n) \qquad (P-6)$$

Where:

TW = Trial Weight in ounces.

RW = The Rotating Weight in pounds.

RPM = The speed of the fan.

In = The distance from the center to the trial weight in inches.

10. Assure the trial weight is securely attached to the blade or hub.

11. Remove the tag and lock and start the fan.

12. Read and record the new vibration amplitude. (This is trial run #1.)

13. Stop the motor and lock out and tag out the motor and secure the blades.
14. Remove the trial weight and attach it to blade number two at the same distance from the center.
15. Repeat steps #10 through #13 for blades number two and three, recording the amplitudes as trial runs two and three respectively.
16. Record the total number of fan blades.
17. Determine the blade angle by dividing 360 by the number of blades.
18. Select a convenient scale to layout the graphical representation of the unbalance by adding the original and the first trial run amplitudes, assure there is adequate room on the paper to lay out a circle with a radius equal to the sum of these amplitudes.

NOTE: Generally, a scale of 1/4 inch equals two mils will be adequate.

19. Draw a circle at the center of the page with a radius equal to the original amplitude, using the selected scale.
20. Draw a line from the center of this circle to the 3 o'clock position on its circumference, and label the line blade one.
21. Using a protractor, measure the blade angle from the blade one line, counterclockwise to locate the point on the circumference for the blade two line.
22. Follow step 21 and locate blade number three.

NOTE: Blade three is located at an angle twice as large as that of blade two.

23. At the location of blade one on the circumference of the original circle, construct a circle with a radius equal to the amplitude of trial run one, using the proper scale factor.
24. At the location of blade two on the circumference of the original circle, construct a circle with a radius equal to the

amplitude of trial run two, using the proper scale factor.

25. At the location of blade three on the circumference of the original circle, construct a circle with a radius equal to the amplitude of trial run three, using the proper scale factor.

26. At the point where the three trial run circles intersect, draw a line to the center of the original circle, and label it T.

27. Measure the length of the line T using the same scale factor, to determine the amplitude.

NOTE: This constructed line represents the net unbalance effect of the trial weight if it were located directly opposite the heavy spot on the actual fan.

28. Determine the correction weight by dividing the original amplitude by the amplitude of T, and multiplying by the amount of trial weight added.

$$\text{CORRECTION WEIGHT} = (O/T) * TW \qquad (P\text{-}7)$$

29. Measure the angle from blade one to the line T. This is the location of the correction weight.

30. If the location of the required correction weight does not fall at a blade location, use one of the following methods to determine correct location and amounts of weights to add.

A. EQUIVALENT WEIGHT METHOD

i. Determine the unbalance by multiplying the distance from the center of the fan to the trial weight location (in inches), by the amount of the trial weight (in ounces).

ii. Measure off the correct angle from blade one and find a suitable location where a weight can be secured. This could be on the fan hub.

iii. Determine the correct amount of weight to be added at this location by dividing the unbalance found in step [i] by the distance from the new location to the center of the fan.

B. VECTOR ANALYSIS METHOD

   i.   On top of the T line, draw a line from the original center with a length equal to the required correction weight.

   ii.   Using the method in step 21, locate the blades on either side of this line.

   iii. At the end of this line, construct two lines parallel to the two blade lines, sufficiently long to pass through the opposite blade line.

   iv. Measure the distance from the center of the original circle, along each of the two blade lines to a point where the parallel lines intersected the blade lines. These distances represent the amounts of correction weights to be added to each of these blades.

31.   Weigh out the proper amount(s) of correction weight(s) to be added. Be sure to compensate for any bolting devices, weldments or holes to be drilled. Securely fasten them at their proper location.

32.   Unlock and remove tag and start the unit, measuring and recording the new vibration amplitude.

33.   Repeat the procedure if amplitudes are not within tolerance.

34.   Lock and tag out motor and remove vibration equipment.

35.   Unlock, untag motor and place unit on-line.

Index

For Product Safety Concerns and Information please contact our EU
representative GPSR@taylorandfrancis.com Taylor & Francis Verlag GmbH
Kaufingerstraße 24, 80331 München, Germany

Printed and bound by CPI Group (UK) Ltd, Croydon, CR0 4YY
01/05/2025
01858490-0002